Electrical and Geometrical Properties of Organic Monolayers

Electrical and Geometrical Properties of Organic Monolayers

Mitsumasa Iwamoto
Tokyo Institute of Technology, Japan

Tetsuya Yamamoto
Hokkaido University, Japan

Zhong-Can Ou-Yang
Chinese Academy of Sciences, China

World Scientific

NEW JERSEY · LONDON · SINGAPORE · BEIJING · SHANGHAI · HONG KONG · TAIPEI · CHENNAI · TOKYO

Published by

World Scientific Publishing Co. Pte. Ltd.

5 Toh Tuck Link, Singapore 596224

USA office: 27 Warren Street, Suite 401-402, Hackensack, NJ 07601

UK office: 57 Shelton Street, Covent Garden, London WC2H 9HE

Library of Congress Cataloging-in-Publication Data

Names: Iwamoto, Mitsumasa, author. | Yamamoto, Tetsuya, author. | Ou-Yang, Zhong-can, author.
Title: Electrical and geometrical properties of organic monolayers /
    Mitsumasa Iwamoto, Tokyo Institute of Technology, Japan, Tetsuya Yamamoto,
    Hokkaido University, Japan, Zhong-can Ou-Yang, Chinese Academy of Science, China.
Description: New Jersey : World Scientific, [2025] | Includes bibliographical references.
Identifiers: LCCN 2024018780 | ISBN 9789814602976 (hardcover) |
    ISBN 9789814602983 (ebook for institutions) | ISBN 9789814602990 (ebook for individuals)
Subjects: LCSH: Monomolecular films--Electric properties. | Organic electronics. |
    Molecular electronics.
Classification: LCC QD509.M65 .I93 2025 | DDC 541/.33--dc23/eng20241030
LC record available at https://lccn.loc.gov/2024018780

British Library Cataloguing-in-Publication Data

A catalogue record for this book is available from the British Library.

For any available supplementary material, please visit
https://www.worldscientific.com/worldscibooks/10.1142/9161#t=suppl

Desk Editors: Kannan Krishnan/Joseph Ang

Typeset by Stallion Press
Email: enquiries@stallionpress.com

Preface

Exploring geometric shapes in nature is a fundamental scientific inquiry. Since pioneering studies on soap bubbles by Plateau and the investigation of interfaces in capillary tubes by Young and Laplace, substantial effort has been dedicated to analyzing the geometric shapes of three-dimensional (3D) materials, including solids, soft matters, liquid crystalline materials, biomaterials, molecules, etc. While many books cover the shapes of 3D materials, the treatment of shapes in two-dimensional (2D) monolayers has been relatively underrepresented. This gap arises due to the unique natures of monolayers, characterized by a symmetric breaking structure at surfaces. Monolayers exhibit spontaneous dielectric polarization properties and liquid-crystalline behaviors, significantly influencing the formation of intriguing domain shapes. Understanding the physical mechanisms stabilizing 2D domains on surfaces requires not only knowledge of material science, including solid-state physics, liquid-crystal physics, and electromagnetism but also proficiency in mathematical techniques capable of addressing geometric problems.

This book aims to introduce an approach to analyze 2D shapes in terms of the electric properties and geometrical structures of organic monolayers. While leveraging geometric mathematical treatments used for 3D materials, this approach incorporates the specific natures of monolayers based on dielectric physics and liquid-crystal physics. This methodology proves particularly beneficial for investigating soft materials, encompassing biomaterials, molecules, and monolayer functions at interfaces. Such knowledge is poised to pave the way for new research fields, where soft materials play pivotal roles in shaping the scientific and technological landscape of the 21st century. Those seeking a profound understanding of the functions and physical mechanisms that stabilize 2D shapes will find inspiration in this book. The content is tailored for a diverse audience, including graduate students undertaking courses in condensed matter physics, soft matter physics, materials science, biophysics, mechanics, electronics, and electrical engineering. It is also beneficial for students enrolled in academic courses like Material Sciences, Condensed Matter Physics, Electronic Materials, and Organic Devices.

This book owes its completion to the enduring collaborative efforts spanning over 25 years between Tokyo and Beijing. Mitsumasa Iwamoto invested considerable time at the Institute of Theoretical Physics (ITP), Beijing, engaging in discussions and studies with Zhong-can, Ou-Yang Reciprocally, O. Z. spent time in Iwamoto's laboratory at Tokyo Institute of Technology (Tokyo Tech) during every new year season. Tetsuya Yamamoto initiated his Ph.D. studies under the supervision of M. I. at Tokyo Institute of Technology, with academic support from O. Z.

We extend our gratitude to the staff, colleagues, visiting scientists, and undergraduate students who contributed to their research and studies in Iwamoto's laboratory at Tokyo Institute of Technology. Special thanks are given to Associate Prof. Dai Taguchi of the Tokyo Institute of Technology, Tokyo, and Prof. Fei Liu of Beihang University, Beijing, for the collaborative work at the Tokyo Institute of Technology.

We warmly welcome comments, criticisms, and suggestions from all readers. Your insights will be greatly appreciated.

Mitsumasa Iwamoto
Emeritus Professor,
Tokyo Institute of Technology, Tokyo Japan

Tetsuya Yamamoto
Associate Professor,
Hokkaido University, Hokkaido Japan

Zhong-can, Ou-Yang
Academician of CAS and Professor,
Institute of Theoretical Physics, Beijing China

About the Authors

Mitsumasa Iwamoto received his B.E., M.E., and D.E. degrees in Electrical Engineering from the Tokyo Institute of Technology (Tokyo Tech), Japan in 1975, 1977, and 1981, respectively. Since 1981, he has been working at Tokyo Tech, and became a Professor in 1995. He is currently an Emeritus Professor at Tokyo Tech (since 2017). Meanwhile, he served as an Academic Icon Professor at the University of Malaya, Malaysia (2013–2016), a Specially Appointed Professor of Tokyo Tech-Tsinghua Univ. Joint Program (Double-degree program) (2017–2022), a Visiting Professor at the China Academy of Science (2017–2020), among other positions.

He is a Distinguished Member and Fellow of the Japan Society of Applied Physics (JSAP), a Fellow of the Institute of Electrical Engineers of Japan (IEEJ), and also a Fellow of the Institute of Electronics, Information and Communication Engineers of Japan (IEICE). His research interests lie in organic materials electronics, particularly, in the development of characterization and evaluation methods for molecular films and organic devices, as well as the study of electrical and optical phenomena of molecular films. He originally developed a Maxwell-displacement-current-measuring method for monolayer studies, established an EFISHG method that can directly visualize electron motions in organic films, among other contributions. Owing to these contributions, he was awarded the JSAP A Award (1991), the IEICE Electronics Society Award (2001), the JSAP M&BE Achievement Award (2015), the IEEJ Ieda Award (2019), and others.

He has organized and chaired international conferences multiple times, such as the International Conference on Nano-Molecular Electronics (ICNME), the International Discussion Conference on Nano interface Controlled Electronic Devices (IDC-NICE), and the International Conference on Molecular Electronics and Bioelectronics (M&BE). On the basis of these conferences, he has edited many special issues in scientific journals such as *Thin Solid Films*, as a guest editor. His previous books published by World Scientific include *Maxwell-Displacement Current and Optical Second Harmonic Generation in Organic Materials* (2021), *Nanoscale Interface for Organic Electronics* (2011), and *The Physical Properties of Organic Monolayers* (2001).

Tetsuya Yamamoto received his B.E. (2003), M.E. (2005), and D.E. (2007) degrees in Physical Electronics from the Tokyo Institute of Technology. He was a postdoctoral fellow at Max Planck Institute for Colloids and Interfaces (2008), Weizmann Institute of Science (2009–2011), University of Strasbourg (2012), and Beihang University (2013–2015). He worked as an Assistant Professor at Tokyo Institute of Technology (2011) and at Nagoya University (2015–2020). He has been working as a specially appointed Associated Professor at the Institute for Chemical Reaction Design and Discovery (ICReDD), Hokkaido University (since 2020). He is an active member of the Molecular Biology Society of Japan and the leader of genome biophysics subgroup in the Biophysical Society of Japan. His research interest is the physical principle in the regulation of gene expression.

Zhong-can Ou-Yang graduated from the Automatics Department of Tsinghua University in 1968, and got his Ph.D. from the Physics Department of Tsinghua University in 1984. From 1987 to 1988 he worked with Prof. W. Helfrich at FU Berlin as an Alexander von Humboldt Foundation fellow. Since 1989, till now, he has been a research faculty of the Institute of Theoretical Physics, Chinese Academy of Sciences (ITP-CAS). From 1998 to 2007, he served as the Director of ITP-CAS.

Zhong-can Ou-yang has made distinguished contributions to a number of theoretical subjects in soft matter physics and biophysics. Based on the Helfrich model, he derived the general shape equation for vesicles of spontaneous curvature and predicted a special type of genus one Clifford torus vesicle with ratio of radii of the generating circles to be $\sqrt{2}$ which was soon confirmed by experiments. He also found the exact and analytic solution of biconcave discoidal shape of human red blood cells which has been a long-standing problem in biology. He proposed a new theory of chiral membrane which explained precisely the experimental observation of low- and high-pitch helical structures formed by tilted chiral lipid bilayers. Later he extended the Helfrich model to study the shapes of carbon nanotubes, the vesicle-to-tube shape transition of bipeptide aggregates, and the icosahedron shape of virus capsid and nano-aggregates. Some of the above-mentioned works were summarized in his *Geometric Method in the Elastic Theory of Membranes in Liquid Crystal Phases,* published by World Scientific (the first edition in 1999 and the second edition in 2018). He also published a booklet *From Soap Bubbles to Biomembrane* (in Chinese) which has been recognized as one of the most popular scientific readings in China.

For his academic achievements, Zhong-can Ou-yang has been awarded several renowned prizes, such as the Achievement in Asia Award of The Overseas Chinese Physics Association in 1993, Chinese Academy of Sciences Awards for Natural Sciences (First class) in 1995, National Natural Science Award (Second class) in 1999, Ho Leung Ho Lee Achievement Prize in 2004, and JSAP Fellow International in 2015. He was elected as the Academician of CAS in 1997 and became a member of the Third World Academy of Sciences in 2003.

Contents

Chapter 1

Introduction

This book treats intriguing domain shapes of organic monolayers, in terms of dielectric and mechanical properties of monolayers. Physical mechanisms that stabilize domains on the water surface are modeled and analyzed using geometric mathematical treatment, with consideration of electrostatic interactions and liquid-crystalline natures in monolayers. This chapter provides background information for studying *electric properties and geometrical structures of organic monolayers*. It begins by introducing Langmuir monomers and their physical and chemical properties in Section 1.1, emphasizing the non-centrosymmetric structures of monolayers at air–water interfaces. Section 1.2 introduces orientational order parameters for quantitatively representing these orientational structures. In Section 1.3, the chapter explores phase transitions and the liquid crystalline nature of monolayers with changes in the density of molecular area by compression. In Section 1.4, we point out that monolayers in the liquid-crystalline phase are an assembly of domains with characteristic shapes, influenced by factors such as electrostatic dipole–dipole interactions. Section 1.5 briefly covers the history of the soft materials shape studies to provide context for the book's focus. Finally, in Section 1.6, the subject of this book is summarized.

1.1.   Organic Monolayers

1.1.1.   *Monolayers on water surfaces*

An organic monolayer consists of organic molecules adsorbed at 2D interfaces, forming a film only one molecular thick. Among such systems, monolayers at the air–water interface, see Fig. 1.1 [1], are the simplest. The conventional methods to form monolayers on water surfaces is to spread a small quantity of amphiphiles (e.g., fatty acids and phospholipids) in a volatile solution such as chloroform by using a micro-pipet. Amphiphiles have hydrophilic "heads" that show attractive interactions with water molecules and hydrophobic "tails" (chains) that show repulsive interactions with water molecules. Amphiphiles are localized at the air–water interface with their hydrophilic groups submerged in the water and their hydrophobic tails left away from the water [1]. Consequently, the structure of the monolayer comprising amphiphiles is non-centrosymmetric with respect to the air–water interface. This fact was discovered by Irving Langmuir (1881–1957) [2] and thus monolayers at air–water interfaces are called Langmuir monolayers.

Molecular films on solid substrates are useful for studying their physical properties. One of elegant film preparation methods is the Langmuir–Blodgett method (LB method), developed by Irving Langmuir and Katharine Blodgett (1898–1979) [3]. A solid substrate is

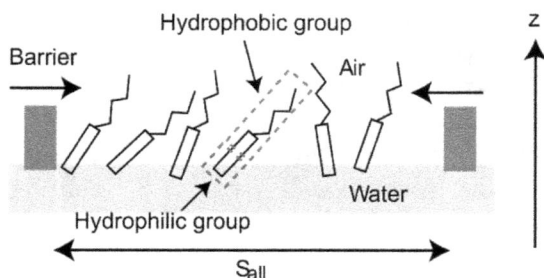

Fig. 1.1:   Amphiphiles have hydrophilic heads and hydrophobic chains and form monolayers at an air–water interface. The thermodynamic states of monolayers are controlled by changing the temperature of the water subphase and/or the area of monolayers. The area of monolayers is controlled by moving the two barriers set up at air–water interfaces; monolayers on the water surface are confined in the space between the two barriers, meanwhile the water beneath is free to move.

vertically moved up and down through the air/water interface, onto which a Langmuir monolayer is prepared. A variety of monolayer films can be formed on water surface, and they are transferred onto the solid substrate layer-by-layer. The molecular films prepared by this method are called Langmuir–Blodgett films (LB films) [4]. The layer-by-layer transfer of Langmuir monolayers onto solid substrates enables us to build up organized artificial multi-layer films, and the resulting deposited film structures are thus attractive for studying the physical properties. The deposited multilayer films on solid substrates are controlled in the precision of molecular length scale, i.e., nano-meter scale, and they are anticipated to work as a source element of molecular functions. As such, the LB technique is an attractive approach to design organized molecular films, and used in a variety of research fields, e.g., physics, chemistry, organic electronics, materials science, bio-engineering, and so on [5]. Noteworthy that there are many other techniques that can deposit monolayers on solid substrates. Among them are self-assembled monolayers (SAMs) that are obtained by merely immersing a solid substrate in a solution of organic molecules with functional groups that can chemically adsorb to the substrates. However, in this book we treat the physics of domain shapes of monolayers, where Langmuir monolayers are the best candidate.

Scientists use monolayers to study the phase transitions and their structures (organic molecules have many degrees of freedom) at 2D interfaces in the context of soft matter physics [1, 6]. For such studies, visualizing monolayers is very effective, where optical microscopic techniques are very often used. Among them are fluorescence microscopes, Brewster angle microscopes, and other state-of-the-art instruments. Experiments using these techniques can reveal the presence of intriguing domains in monolayers on water surfaces. Accordingly, physicists and chemists show interest in the physical mechanisms that stabilize the characteristic 2D domain shapes, which are very different from the shapes of 3D liquid materials.

1.1.2. *Bilayer membranes and monolayer*

Lipids in an aqueous solution spontaneously assemble a lipid bilayer membrane, where the hydrophobic chains of lipids in two monolayers are juxtaposed to shield these chains from water molecules by hydrophilic heads of these lipids. Cell membranes are composed of lipid bilayer membranes, in which membrane proteins can freely diffuse due to the fluidity of the membranes. The model called fluid mosaic model was proposed by Singer and Nicolson [7], and is well adopted as the basis of membrane biophysics.

Monolayers composed of lipids are used as model systems to study the conformations of proteins in the membranes and the interactions between proteins and lipids. Lipid monolayers are relatively simple systems among the models of cell membranes and many experimental techniques are applied to characterize lipid monolayers. However, there are a couple of differences between lipid monolayers and bilayer membranes: Bilayer membranes show interactions between their two monolayers (leaflets). In many model bilayer membranes, the structure of the constituting two monolayers is symmetric, whereas monolayers show non-centrosymmetric orientational structures with respect to the water surface. Nevertheless, lipid monolayers show many properties that are common to bilayer membranes, and their physical properties are extensively studied. Indeed, the two-monolayer systems of cell membranes are believed to be asymmetric. LB technique enables us to prepare asymmetric bilayer membranes and thus is used to study the effect of membrane asymmetry.

Membrane biologists have proposed a so-called raft model that states that stable nanoscopic lipid domains in biological membranes play an important role in several biological processes. However, because these lipid domains are smaller than the diffraction limit of optical microscopes and biological membranes include many types of lipids and proteins, domains in lipid monolayers have been studied as model systems to study the physical mechanisms that stabilize lipid rafts. Their shapes and dynamics are observed to extract the information of "line tensions" that are believed to be key physical quantities relevant to the stability of lipid rafts. Understanding the physical mechanisms involved in the shapes of domains in monolayers is important. It is thus of interest to elucidate the physical mechanisms that govern the formation of characteristic-shaped domains in monolayers, in terms of bilayers of lipids.

1.1.3. *Organic monolayers in devices*

The information on orientational structure of organic monolayers at interfaces is useful in other research fields, e.g., in organic electronics [5], where understanding nano-scale electrostatics generated at these interfaces and preparation of well-defined and well-organized mono- and multi-layer films are very important to make use well of organic molecular functions in organic devices. Consequently, deep understanding of the physics of 2D state of organic monolayers at interfaces is necessary [8,9]. Monolayers on water surface are ideal model systems, and give insightful information on nano-scale electronics, including nano-electrochemistry.

As briefly mentioned in Section 1.1, the physics of monolayers is important in a variety of research fields, including science, chemistry, biology, and electronics.

1.2. Orientational Order Parameters

Monolayers at interfaces show non-centrosymmetric orientational structures due to symmetry breaking. As we will discuss extensively in this book, the dielectric spontaneous polarizations of monolayers, caused by non-centrosymmetric orientational structure, play an important role in stabilizing the shapes of domains in monolayers on water surfaces. In this situation, we encounter a basic question of how we should quantitatively represent non-centrosymmetric orientational structures. To formalize the dielectric physics of monolayers, it is convenient to introduce an idea of using *orientational order parameters* [10–12]. This is basically an extension of the idea employed in the field of physics of liquid crystals. In Chapter 2, we will use these orientational order parameters to construct the dielectric physics of monolayers. In the following subsections, we briefly summarize the basis of the *orientational order parameters*.

1.2.1. *Orientational order parameter of nematic liquid crystals*

Nematic liquid crystals are typical bulk liquid materials that show orientational ordering. The simplest and most relevant examples of nematic liquid crystals are those composed of rod-shaped molecules such as 4-Cyano-4′-pentylbiphenyl (5CB) molecules. The orientations of rod-shaped molecules are defined by the directions of their long-axes, see Fig. 1.2. Nematic liquid crystals composed of rod-shaped molecules show orientational ordering toward a direction that is

Isotropic liquids   Liquid crystals   Monolayers

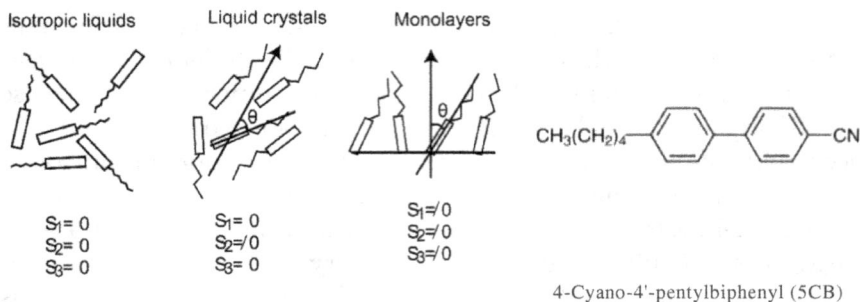

$S_1 = 0$ $S_1 = 0$ $S_1 \neq 0$
$S_2 = 0$ $S_2 \neq 0$ $S_2 \neq 0$
$S_3 = 0$ $S_3 = 0$ $S_3 \neq 0$

$CH_3(CH_2)_4$—⟨ ⟩—⟨ ⟩—CN

4-Cyano-4'-pentylbiphenyl (5CB)

Fig. 1.2: The orientational order parameters S_n are defined in Eq. (1.3). The lowest order terms S_1, S_2, and S_3 are useful to represent state of matters: Isotropic liquids have $S_1 = 0$, $S_2 = 0$, and $S_3 = 0$. Liquid crystals have $S_1 = 0$, $S_2 \neq 0$, and $S_3 = 0$. The non-centrosymmetric orientational structure of monolayers is characterized by $S_1 \neq 0$, $S_2 \neq 0$, and $S_3 \neq 0$. 4-Cyano-4'-pentylbiphenyl (5CB) molecule is a typical rod-shaped molecule.

specified by a unit vector **m** called director (or orientational vector). In nematic liquid crystals, the number of molecules that orient toward the director $+\mathbf{m}$ is, on average, the same as the number of molecules that orient toward the opposite direction, $-\mathbf{m}$, even when the molecular structures (atomic arrangements) of constituent molecules themselves are non-centrosymmetric, see Fig. 1.2. In a mean field picture, the extent of orientational ordering of rod-shaped molecules in nematic liquid crystals is characterized by using the orientational order parameter [10–12]

$$S_2 = \left\langle \frac{3\cos^2\theta - 1}{2} \right\rangle, \tag{1.1}$$

where θ is the tilt angle between the orientation of a rod-shaped molecule and the director **m**, see Fig. 1.2. The angular bracket $\langle\ \rangle$ stands for the thermodynamic average

$$\langle A(\theta, \phi) \rangle = \int d(\cos\theta) \int d\phi A(\theta, \phi) f(\theta, \phi), \tag{1.2}$$

where $f(\theta, \phi)$ is the orientational distribution function with the tilt angle θ and the tilt azimuth ϕ that represent the direction of the projection of the long-axis of a rod-shaped molecule onto the normal plane of the director. The orientational order parameter S_2 of nematic liquid crystals is non-zero, whereas the orientational order parameter S_2 of isotropic liquids is zero, see Fig. 1.2. The orientational order parameter S_2 characterizes the orientational order of nematic liquid crystals. Indeed, many of the characteristic physical properties of nematic liquid crystals are related to the non-zero-orientational order parameter S_2 [11].

Mathematically, the orientational order parameter S_2 is defined using the thermodynamic average $\langle \ \rangle$ of the Legendre polynomials of the second order, $\langle (3\cos^2\theta - 1)/2 \rangle$ (see Eq.(1.1)). The orientational-order parameter S_2 represents the extent that (the long-axes of) constituent rod-shaped molecules orient toward director. Because it is an even function of $\cos\theta$, the molecules, whose long-axes are directed in one direction, contribute to the order parameter in the same way as the molecules, whose long-axes are directed in the opposite direction, see Fig. 1.2. This property manifests the centrosymmetric orientational structure of nematic liquid crystal.

Many types of organic molecules, e.g., disk-shaped molecules, banana-shaped molecules, etc., exhibit nematic liquid crystal properties. The extension of the treatment of rod-shaped molecules to biaxially molecules, chiral molecules, and other complex-shaped molecule cases is possible, though we need to use more generalized form of order parameters (Ref. [10]).

1.2.2.   *Orientational order parameters of monolayers*

The non-centrosymmetric orientational structure is characteristic of monolayers at air–water interface. This non-centrosymmetric structure results from the amphiphilic nature of constituent organic molecules (symmetry breaking), see Fig. 1.1. If we define unit vectors from the hydrophilic head of a constituent molecule toward the end of the hydrophobic chain of the molecule, the unit vector is directed toward the air. The extent that these vectors are directed toward the normal of the air–water interface, represents the extent of the orientational ordering of constituent molecules of monolayers. The orientational order parameter S_2 is not sufficient to characterize the non-centrosymmetric orientational structures of monolayers on water surface.

Mathematically, Legendre polynomials of the odd-number-th order are odd functions of $\cos\theta$. The order parameters defined for these polynomials enable us to quantitatively characterize the non-centrosymmetric orientational structures of monolayers, see Fig. 1.2. We thus introduce a set of orientational order parameters [10, 13] defined by the form

$$S_n = \langle P_n(\cos\theta) \rangle \quad (n = 1, 2, 3, \ldots). \tag{1.3}$$

S_n is an extension form of the nematic order parameter S_2 to the Legendre polynomials $P_n(\cos\theta)$ of n-th order, $n = 1, 2, \ldots$. We here treat a monolayer that shows orientational order toward the monolayer normal $\mathbf{n}$ and thus the director $\mathbf{m}$ of these monolayers is parallel to the monolayer normal, see Fig. 1.2.

In this book, we mainly treat the orientational order parameters of the first three lowest orders, S_1, S_2, and S_3 that have the forms

$$S_1 = \langle \cos\theta \rangle, \tag{1.4}$$

$$S_3 = \left\langle \frac{5\cos^2\theta - 3\cos\theta}{2} \right\rangle, \tag{1.5}$$

and S_2 (see Eq. (1.1) for the expression of S_2). The orientational order parameters S_1 and S_3 can characterize the non-centrosymmetric orientational structures of monolayers, whereas S_2 represents the orientational order in a similar manner to nematic liquid crystals. One can also think that the orientational order parameter S_1 is the average of the cosine of the tilt angle of the long-axes of constituent molecules, whereas S_2 is related to the variance of the orientation of constituent molecules.

1.2.3.   *Classification of liquid matters by orientational order parameters*

The orientational order parameters S_1, S_2, and S_3 are useful parameters to characterize the orientational order of liquid matters comprising rod-like molecules: The orientation of molecules in isotropic liquids is random and the orientational structures of these isotropic liquids are represented by $S_1 = 0$, $S_2 = 0$, and $S_3 = 0$ (see Fig. 1.2). Rod-shaped molecules in nematic liquid crystals show centrosymmetric orientational structures, which are characterized by $S_1 = 0$, $S_2 \neq 0$, and $S_3 = 0$. Monolayers show the non-centrosymmetric orientational structures, which are characterized by $S_1 \neq 0$, $S_2 \neq 0$, and $S_3 \neq 0$. Monolayers are therefore the most general systems from viewpoint of order parameters. The parameters S_1 and S_2 have simple physical meanings; the average orientation toward the monolayer normal **n** and its variance, respectively. What is the physical meaning of other orientational order parameters of higher order? Because of the mathematical properties of the Legendre polynomials, orientational distribution functions are expanded with respect to the Legendre polynomials in the form

$$f(\theta, \phi) = \frac{1}{2\pi} \sum_{n=0}^{\infty} \frac{2n+1}{2} S_n P_n(\cos \theta). \qquad (1.6)$$

This is shown by substituting Eq. (1.6) into Eq. (1.3) and using the fact that the Legendre polynomials are orthogonal functions. The orientational order parameters S_n are thus the expansion coefficients of orientational distribution functions; the orientational distribution function is reproduced, once a set of orientational order parameters S_n ($n = 1, 2, 3, \ldots$) are determined. Equation (1.6) is also useful to derive the physical quantities, e.g., dielectric polarizations, as functions of the orientational order parameters S_n. Noteworthy that in many cases, using the first three order parameters, i.e., S_1, S_2, and S_3, are sufficient to discuss the physical properties of monolayers. For example, the spontaneous polarization of monolayer can be described using order parameter S_1 [13, 14].

1.3. Phases and Liquid Crystalline Nature of Monolayers

Since the discovery of the influence of impurities on the surface tension by Agnes Pockels (1862–1935) [15], the fact that the monolayer at an air–water interface decreases the surface tension has been well accepted by many scientists. Surface pressure, the decrement of the surface tension, has been routinely characterized to understand the thermodynamic properties of monolayers. The surface pressure Π of monolayers is typically measured by changing the area per molecule A of the monolayer at a constant temperature. Such measurements are thus called $\Pi - A$ isotherm measurement. Measuring $\Pi - A$ isotherm at various temperatures provides the information of

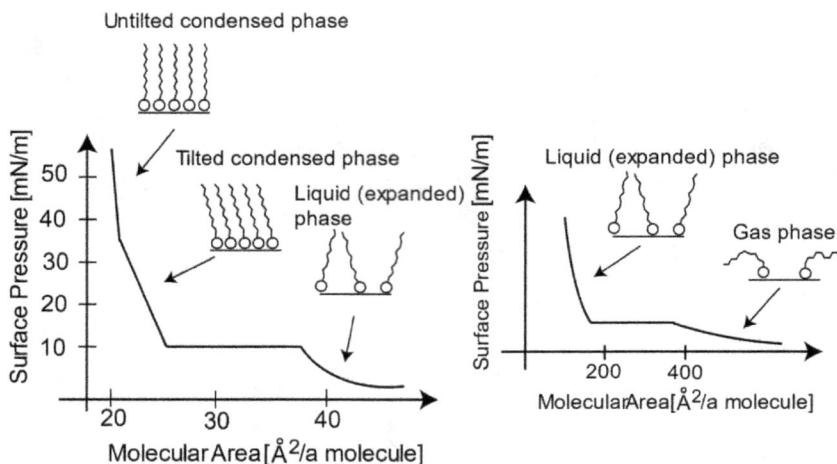

Fig. 1.3: Surface pressure of monolayers as a function of area per molecule. Monolayers are in gaseous phase when area per lipid is very large (inset) and experience phase transitions to liquid expanded phase and then condensed phase. There is a two-phase coexistent state between liquid expanded phase and condensed phase. This figure is reproduced from Ref. [6].

the state of monolayers, where the plateaus (in some cases, quasi-plateaus) indicate their phase transitions (see Fig. 1.3).

In the limit of large area per molecule A, a monolayer is in the gaseous phase, where the surface pressure is mainly due to the translational entropy of constituent molecules. By decreasing the area per molecule, the monolayer shows the phase transition to "liquid expanded phase", which is the 2D analogue of isotropic liquids. In this phase, the long-axes of constituent molecules show orientational ordering toward the normal to the interface, but are isotropic with respect to the tilting azimuth. With further decrease of area per molecule A, the monolayer shows the phase transition to "liquid condensed phase". There are two classes of liquid condensed phase — the *tilted* phase and the *untilted* phase. In the *tilted* phase, the constituent molecules show not only the orientational ordering in the normal to the interface but also the orientational ordering with respect to the tilting direction of the long-axes of constituent molecules — *in-plane orientational ordering*. The in-plane orientational ordering results from the molecular interactions between constituent molecules. This aspect of the *in-plane* orientational ordering of monolayers is similar to nematic liquid crystals, and monolayers indeed show liquid crystalline properties in this phase. The physical properties due to the liquid crystalline nature of monolayers is one of important subjects of this book. In the *untilted* phase, constituent molecules are closely packed so that the long-axes of constituent molecules cannot tilt toward the water surface. With further decrease of the area per molecule, monolayers become unstable and collapse to multilayers. Experiments on fatty acid monolayers by using the state-of-the-art techniques have revealed that the liquid condense phase is indeed subdivided into various subphases, and these subphases were reviewed by many scientists, e.g., in the article by Kaganer *et al.* [6].

1.4.   Shapes of Domains in Monolayers

During the phase transition, monolayers assemble domains of liquid condense phase that coexist with domains of liquid expanded phase. Microscopic visualization techniques, such as fluorescence microscopy [16] and Brewster angle microscopy [17, 18], have been developed to visualize domains. Domains of an isotropic liquid phase that coexist with domains of another isotropic liquid phase are expected to be spherical due to the surface tension. In contrast to this situation, the shapes of domains in monolayers show a variety of shapes that are quite different from circular, which is expected from naïve guesses from the domains of 3D isotropic liquids. The shapes of domains depend on the thermodynamic state and the chemistry of constituent molecules. For example, monolayers of dipalmitoyl-phosphatidylcholine (DPPC), which is a typical saturated lipid, show clover-shaped domains (see Fig. 1.4(a) and (b)). DPPC is a chiral molecule; the molecular structure of one optical isomer, L-DPPC, is the mirror image of the molecular structure of the other, D-DPPC. Indeed, the lobes of the clover-shaped domains in L-DPPC monolayers are bent in the opposite direction to the lobes of the clover-shaped domains in D-DPPC monolayers; the shape of domains in L-DPPC monolayers is the mirror image of the shape of domains in D-DPPC monolayers (see Fig. 1.4(a) and (b)). What is the physical mechanism involved in the formation of the clover-shaped domains in DPPC monolayers?

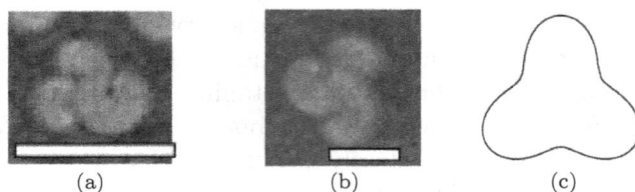

(a)   (b)   (c)

Fig. 1.4:   Domains in L-DPPC (a) and D-DPPC (b) monolayers (where the surface pressure was adjusted to 12.4 mN/m and 10.3 mN/m, respectively). These domains stabilize clover-shaped domains and their lobes are bent toward anticlockwise for the case of L-DPPC and clockwise direction for the case of D-DPPC. The scale bars indicate 30 μm. McConnell's theory can predict that monolayers stabilize clover-shaped domains, but their lobes are straight ones (c).

Amphiphilic molecules have permanent electric dipoles along the long-axes of these molecules. Because of the non-centrosymmetric orientational structures of monolayers, the electric dipoles are oriented in the direction normal to the air–water interface and the constituent molecules thus show repulsive dipole–dipole interactions. The dipole–dipole interactions contribute to the free energy as the electrostatic energy. In 3D liquid materials, domains of isotropic liquids in another isotropic liquids grow by coarsening or coalescence to minimize the surface energy because the free energy of this system is governed by the surface energy. Andelman *et al.* [19] and McConnell [20] considered the electrostatic energy due to the dipole–dipole interactions to predict that domains in monolayers have stable size, which is determined by the balance between the surface energy and the electrostatic energy. Indeed, McConnell's theory predicted that the clover-shaped domains are stabilized by the balance between the surface energy and the electrostatic energy. The discovery that the electrostatic dipole–dipole interactions play an important role in the domain pattern of monolayers is a significant step toward the understanding of the geometrical properties of monolayers. However, it does not provide a complete explanation. For example, McConnell's theory communicates the limitation in predicting the dependence of the bending direction of the lobes of domains on the chirality of constituent molecules (see Fig. 1.4(c)), as experimentally observed in DPPC monolayers. The liquid crystalline nature of monolayers in the liquid condensed phase might be one of the missing pieces for analyzing domain shapes, besides the molecular structure of constituent molecules.

1.5. History of the Study on Shapes of Soft Materials

It is instructive here to go back to the history of the geometric study on shapes of soft materials, to make clear the standpoint of this book. Shape problems stem from real interfaces of crystals, liquids, and biological systems in nature. The pioneering studies were soap bubbles by Plateau [21] and the shape of interfaces in a capillary tube by Young [22] and Laplace [23], and these studies were done in the 19th century. Minimal surfaces were analyzed as the variation problem $\delta \oint dA = 0$ and surfaces with constant curvature, $\delta[\Delta P \int dV + \lambda \oint dA] = 0$, where ΔP is the pressure between air/liquid interface, λ is the surface tension, A is the area of the interface, and V is the volume of liquid phases. This approach was very successful, and much effort has been devoted to analyzing geometric shapes of 3D materials. Nowadays, such kind of variations have been developed to include the integral of the mean curvature H and Gaussian curvature K of the interface, for instance, the important variation equation of the Helfrich free energy $\delta F_H = 0$ [24, 25] has been well adopted to account for intriguing shapes of 3D materials in nature [26]. However, shapes of two-dimensional (2D) domains in monolayers have not been well treated, possibly owing to the specific physicochemical monolayer properties as mentioned in Sections 1.2–1.4.

1.6. Subject of this Book

As discussed in Section 1.4, the electrostatic dipole–dipole interactions play an important role in the shapes of domains in monolayers, however, there are missing pieces to fully understand the physical mechanisms involved in the shapes of domains. One of the missing pieces may be the liquid crystalline nature of monolayers in liquid condensed phase, where constituent molecules show orientational ordering with respect to the tilting directions of their long-axes. Of course, actual molecular structures and their properties might be other missing pieces.

Electric dipoles are origin of the dielectric polarization, which is a key quantity in the dielectric physics. In this book, we extend the dielectric physics of 3D solids to that of 2D monolayers, by considering their liquid crystalline natures and molecular structures, and formulate equations that lead to the solutions of stable shapes of domains in monolayers. The repulsive electrostatic dipole–dipole interactions result from the non-centrosymmetric orientational structures of monolayers and the electrostatic energy due to the interactions should be a function of the orientational order parameters S_1, S_2 and S_3 etc. In Chapter 2, we review the electrostatics of isotropic 3D materials, and then derive the dielectric polarization generated in 2D monolayers by extending the dielectric physics of isotropic 3D materials. This chapter makes the basis of this book. In Chapter 3, experimental techniques to measure dielectric polarizations in monolayers are reviewed, and show a way for determining orientational order parameters that characterize monolayers, and so on. In Chapter 4, using experimental results of monolayers, we discuss the liquid crystalline nature of monolayers from various viewpoints. In Chapter 5, we show a way to analyze domain shapes by using the geometric mathematical treatment, i.e., *differential geometry*, and derive shape equations. Minimizing the free energy of monolayers leads to shape equations, and then typical solutions are shown. In Chapter 6, we further pay attention to the physical meaning of the minimum of free energy in terms of the Maxwell stress. The *in-plane* polar orientational order in monolayers is treated in terms of liquid-crystalline nature of the monolayers, and formation of cusp domains is discussed. In Chapter 7, we show a comprehensive approach for analyzing domain shapes, by more paying attention to actual molecular structures, where the effect of both polar property and liquid-crystalline property are included together in free energy of monolayers.

References

[1] G. L. Gaines, *Insoluble Monolayers at the Liquid–Gas Interface*, Wiley-Interscience, New York, 1966.

[2] I. Langmuir, The constitution and fundamental properties of solids and liquids II. Liquids, *J. Am. Chem. Soc.*, 39, 1848 (1917).

[3] K. B. Blodgett and I. Langmuir, Built-up films of barium stearate and their optical properties, *Phys. Rev.*, 51, 964 (1937); K. B. Blodgett, Films built by depositing successive monomolecular layers on a solid surface, *J. Am. Chem. Soc.*, 57, 6, 1007–1022 (1935).

[4] G. G. Roberts and C. W. Pitt (ed.), *Langmuir-Blodgett Films, 1982: Proceedings of the First International Conference on Langmuir-Blodgett Films*, Durham, Gt. Britain, September 20–22, 1982, Thin Solid Films, Elsevier Scientific Pub., 1983.

[5] G. Roberts (ed.), *Langmuir-Blodgett Films*, Springer, Berlin, 1990; A. Ulman, *An Introduction to Organic Thin Films*, Academic Press, Cambridge, 1991.

[6] V. M. Kaganer, H. Möhwald, and P. Dutta, Structure and phase transitions in Langmuir monolayers, *Rev. Mod. Phys.*, 71, 779 (1999).

[7] S. J. Singer and G. L. Nicolson, The fluid mosaic model of the structure of cell membranes, *Science*, 175, 720–731 (1972).

[8] W. R. Salaneck, K. Seki, A. Kahn, and J. J. Pireaux (eds.), *Conjugated Polymers and Molecular Interfaces*, Marcell Dekker, Inc., New York (2002).

[9] M. Iwamoto, Y. Kwon, and T. Lee (eds.), *Nanoscale Interface for Organic Electronics*, World Scientific, New Jersey (2011).

[10] G. R. Luckhurst and G. W. Gray, *The Molecular Physics of Liquid Crystals*, Chapter 3, Academic Press, London, 1979.

[11] S. Chandrasekhar, *Liquid Crystals*, 2nd ed., Cambridge University Press, New York, 1992.

[12] P. G. de Gennes and J. Prost, *The Physical Properties of Liquid Crystals*, Clarendon Press, Oxford, 1993.

[13] M. Iwamoto and C. X. Wu, *The Physical Properties of Organic Monolayers*, World Scientific Publishing Ltd., Singapore, 2001.

[14] M. Iwamoto and D. Taguchi, *Maxwell Displacement Current and Optical Second-Harmonic Generation in Organic Materials: Analysis and Application for Organic Electronics*, World Scientific Publishing Ltd., Singapore, 2021.

[15] A. Pockels, Surface tension, *Nature*, 46, 437 (1891).

[16] V. T. Moy, D. J. Keller, H. E. Gaub, and H. M. McConnell, Long-range molecular orientational order in monolayer solid domains of phospholipid, *J. Phys. Chem.*, 90, 3198 (1986).

[17] D. Hönig and D. Möbius, Direct visualization of monolayers at the air–water interface by Brewster angle microscopy, *J. Phys. Chem.*, 95, 4590–4592 (1991).

[18] S. Henon and J. Meunier, Microscope at the Brewster angle: Direct observation of first-order phase transitions in monolayers, *Rev. Sci. Instrum.*, 62(4), 936 (1991).

[19] D. Andelman, F. Brochard, and J. F. Joanny, Phase transitions in Langmuir monolayers of polar molecules, *J. Chem. Phys.*, 86, 3673–3681 (1987).

[20] H. M. McConnell and V. T. Moy, Shapes of finite two-dimensional lipid domains, *J. Phys. Chem.*, 92, 4520–4525 (1988).

[21] J. Plateau, *Statique Experimentable et Theoretique des Liquides Soumis aux Seules Forces Moleculaires*, Gauthier Villars, Paris, 1873.

[22] T. Young, An essay on the cohesion of fluids, *Philos. Trans. R. Soc.*, 95, 65 (1805).

[23] P. S. Laplace, *Traite de Mecanique Celeste*, Gauthier Villars, Paris, 1839.

[24] W. Helfrich, Elastic properties of lipid bilayers: Theory and possible experiments., *Z. Naturforsch. C*, 28(11), 693–703 (1973).

[25] Z. C. Ou-Yang and W. Helfrich, Instability and deformation of a spherical vesicle by pressure, *Phys. Rev. Lett.*, 59, 2486 (1987).

[26] Z. C. Ou-Yang, J. X. Liu, and Y. Z. Xie, *Geometric Methods in the Elastic Theory of Membranes in Liquid Crystal Phases*, World Scientific, Singapore, 1999.

Chapter 2

Dielectric Physics of Monolayers; Theory of Dielectric Polarization of Monolayers

Dielectric properties of monolayers on solid and liquid surfaces exhibit notable distinctions from those of three-dimensional (3D) bulk materials. In this chapter, the dielectric properties of monolayers on solid and liquid surfaces are examined, focusing on their differences from those of 3D bulk materials. Section 2.1 discusses the microscopic electrostatics of general solids based on the general theory of electromagnetism. Section 2.2 provides a brief description of the microscopic mechanisms of dielectric polarizations in isotropic materials. In Section 2.3, physical quantities describing the dielectric nature of monolayers are formulated using orientational order parameters. This includes spontaneous polarizations induced by orientational ordering, electrostatic energies due to spontaneous polarization, Maxwell stress caused by spontaneous polarization, and other relevant factors.

2.1. Macroscopic Electrostatics of Matters

2.1.1. *Fundamental laws of microscopic electrostatics*

Here, we begin our discussion on the macroscopic electrostatics of general dielectric materials, based on the general theory of electromagnetism [1–3]. In electrostatics, (static) electric fields

$\mathbf{e}\,[N/C = V/m]$ are determined by the fundamental laws of electrostatics;

$$\nabla \cdot \mathbf{e}(\mathbf{r}) = \frac{\rho(\mathbf{r})}{\epsilon_0}, \tag{2.1}$$

$$\nabla \times \mathbf{e}(\mathbf{r}) = 0, \tag{2.2}$$

where ∇ is gradient of 3D spaces; $\nabla \cdot \mathbf{e}$ is the divergence of vector fields $\mathbf{e}(\mathbf{r})$, and $\nabla \times \mathbf{e}$ is the rotation of vector fields $\mathbf{e}(\mathbf{r})$. $\mathbf{r}$ is positional vector in 3D space. $\rho(\mathbf{r})$ $[C/m^3]$ is the volume density of electric charges, and $\epsilon_0 (= 8.85416 \times 10^{-12}$ F/m) is the dielectric permittivity of vacuum. Equation (2.2) suggests that electric fields are irrotational fields. Electric fields are thus represented by using (the negative of) the gradient of a scaler function of electrostatic potentials $v(\mathbf{r})$ $[V$ $(= \mathrm{Nm/C})]$;

$$\mathbf{e}(\mathbf{r}) = -\nabla v(\mathbf{r}). \tag{2.3}$$

The electric fields $\mathbf{e}$ are uniquely determined using Eqs. (2.1) and (2.2) with boundary conditions that are given by the electrostatic potentials at the boundary or the components of electric fields that are normal to the boundary. Equations (2.1) and (2.3) are combined to the Poisson equation that has the form

$$\nabla^2 v(\mathbf{r}) = -\frac{\rho(\mathbf{r})}{\epsilon_0}. \tag{2.4}$$

For the cases of $\rho(\mathbf{r}) = 0$, Eq. (2.4) is called Laplace equation;

$$\nabla^2 v(\mathbf{r}) = 0. \tag{2.5}$$

Laplace equation determines electric fields in the system formed by externally applied electric fields (or, more precisely, electric fields generated from electric charges that are located at the exterior of the system). Electric fields that are generated from electric charges in the system are called *Poisson fields*, whereas those arising from external applied electric fields are called *Laplace fields*.

Dielectric polarizations originate from microscopic electric dipoles. It is thus important to know electric fields generated from electric dipoles, e.g., to discuss the effect of alignment of dipolar molecules. To this end, we use a mathematical relationship

$$\nabla^2 \frac{1}{r} = -4\pi\delta(\mathbf{r}), \tag{2.6}$$

where $r(=|\mathbf{r}|)$ is the length of positional vectors $\mathbf{r}$ (here and after, $|\mathbf{A}|\,(=A)$ indicates the magnitudes of vectors $\mathbf{A}$) and $\delta(\mathbf{r})$ is Dirac's delta function. The volume integral of this function is 1 (unity), as long as this volume integral includes $\mathbf{r} = \mathbf{0}$;

$$\int_{\mathbf{r}=0\in V} dV\,\delta(\mathbf{r}) = 1. \tag{2.7}$$

Comparing Eqs. (2.6) and (2.5) leads to electrostatic potentials generated by a point charge, $\rho(\mathbf{r}) = q\delta(\mathbf{r})$, located at the origin $\mathbf{0}$ in the form

$$v(\mathbf{r}) = \frac{1}{4\pi\epsilon_0}\frac{q}{r}. \tag{2.8}$$

The electrostatic potential generated by an electric dipole, which is a pair of a positive charge $+q$ at $\mathbf{r} = \mathbf{d}/2$nd a negative charge $-q$ at $\mathbf{r} = -\mathbf{d}/2$, is thus calculated in the form

$$v_{\text{dip}}(\mathbf{r}) = \frac{1}{4\pi\epsilon_0}\left[\frac{q}{\left|\mathbf{r}-\frac{\mathbf{d}}{2}\right|} - \frac{q}{\left|\mathbf{r}+\frac{\mathbf{d}}{2}\right|}\right] \cong \frac{1}{4\pi\epsilon_0}\frac{\boldsymbol{\mu}\cdot\mathbf{r}}{r^3}, \tag{2.9}$$

where $\boldsymbol{\mu}$ is an electric dipole. The magnitude of the electric dipole $|\boldsymbol{\mu}|$ is called dipole moment and given by $\mu = qd$ [Cm]. Note that μ is often treated using the unit of Debye D ($1\text{D} = 3.33564 \times 10^{-30}$ Cm) [4]. This unit is conventional when we treat permanent dipoles in real polar molecules in solids. The last equation of Eq. (2.9) is derived by expanding the second equation with respect to $|\mathbf{d}/r|$ and omitting higher-order terms. The ratio of the higher-order terms to the term shown in the last equation of Eq. (2.9) is $|\mathbf{d}|^2/(4r^2)P_3(\cos\theta)/P_1(\cos\theta)$, where $P_n(\cos\theta)$ is the Legendre polynomials of nth order and θ is the angle between positional vectors $\mathbf{r}$ and dipole moment $\boldsymbol{\mu}$; Eq. (2.9) is a good approximation for $r \le |\mathbf{d}/2|$ Substituting Eq. (2.9) into Eq. (2.3) leads to electric fields in the form

$$\mathbf{e}_{\text{dip}}(\mathbf{r}) = \frac{1}{4\pi\epsilon_0}\left[\frac{3\boldsymbol{\mu}\cdot\mathbf{r}}{r^5}\mathbf{r} - \frac{\boldsymbol{\mu}}{r^3}\right]. \tag{2.10}$$

For the cases of electric permanent dipoles, whose magnitude has no relation with external applied electric fields $\mathbf{e}(\mathbf{r})$, an electric dipole $\boldsymbol{\mu}$ located at $\mathbf{r}$ stores electrostatic energy that has the form

$$W = q\phi\left(\mathbf{r} + \frac{\mathbf{d}}{2}\right) - q\phi\left(\mathbf{r} - \frac{\mathbf{d}}{2}\right) \simeq -\boldsymbol{\mu} \cdot \mathbf{e}(\mathbf{r}). \tag{2.11}$$

The second equation has the form of electrostatic energy arising from a positive charge $+q$ at $\mathbf{r} + \mathbf{d}/2$nd a negative charge $-q$ at $\mathbf{r} - \mathbf{d}/2$ that consists of the electric dipole $\boldsymbol{\mu}$.

Electrostatic energy arising from the interactions between an electric dipole $\boldsymbol{\mu}_i$ at $\mathbf{0}$ and an electric dipole $\boldsymbol{\mu}_j$ at $\mathbf{r}$ thus has the form

$$W = \frac{1}{4\pi\epsilon_0}\left[\frac{\boldsymbol{\mu}_i \cdot \boldsymbol{\mu}_j}{r^3} - \frac{3(\boldsymbol{\mu}_i \cdot \mathbf{r})(\boldsymbol{\mu}_j \cdot \mathbf{r})}{r^5}\right]. \tag{2.12}$$

Here Eq. (2.10) is used for electric fields $\mathbf{e}(\mathbf{r})$ in Eq. (2.11).

The concept of electric dipoles, i.e., a pair of positive and negative charges, is generalized to arbitrary charge distribution systems of positive and negative charges. It is defined using their first positional moment:

$$\boldsymbol{\mu} = \int_V dV \mathbf{r}\rho(\mathbf{r}) + \int_S dS \mathbf{r}\sigma(\mathbf{r}), \tag{2.13}$$

where $\rho(\mathbf{r})$ is the volume density of electric charges in the bulk of an arbitrary volume V and $\sigma(\mathbf{r})$ is the area density of electric charges at the closed surfaces that enclose this volume. Equation (2.13) returns to $\mu = qd$ when $\rho(r)$ is given by a pair of positive and negative charges in the way as $\rho(r) = q\delta(\mathbf{r} - \mathbf{d}/\mathbf{2}) - q\delta(\mathbf{r} + \mathbf{d}/\mathbf{2})$ and $\sigma(\mathbf{r}) = 0$.

2.1.2.   Fundamental laws of macroscopic electrostatics

The theory of electrostatics treats ideal dielectric materials as a continuum medium. While dielectric materials are composed of atoms and molecules at the microscopic level, the continuum theory deals with their properties at a length scale larger than the molecular length scale. Therefore, the continuum theory of dielectric materials is referred to as macroscopic electrostatics. The continuum theory of dielectric materials is thus called macroscopic electrostatics. Noteworthy that (microscopic) electric fields $\mathbf{e}(\mathbf{r})$ in real dielectric materials change steeply in the molecular length scale, but these molecular scale gradients of electric fields does not influence the dielectric properties at larger length scale. In the continuum theory of dielectric materials, electric fields (and electrostatic potentials) are averaged over a macroscopic length scale;

$$\mathbf{E}(\mathbf{r}_0) = \frac{1}{\Delta V} \int_{\Delta V} \mathrm{d}V \mathbf{e}(\mathbf{r}_0 + \mathbf{r}), \tag{2.14}$$

$$\phi(\mathbf{r}_0) = \frac{1}{\Delta V} \int_{\Delta V} \mathrm{d}V v(\mathbf{r}_0 + \mathbf{r}), \tag{2.15}$$

where the volume integral should be performed with respect to $\mathbf{r}$ for a volume ΔV that is larger than the molecular length scale, but is smaller than the system size. This treatment eliminates the steeply changes of electric fields at the length scales that are smaller than the size of volume elements ΔV; macroscopic electric fields are uniform within a volume element ΔV to a good approximation. With this treatment, positional vector $\mathbf{r}$ points to a volume element that includes many molecules. The average of the thermodynamic quantities of molecules over a volume element ΔV is thus equivalent to thermodynamic average.

Let us here consider an ideal dielectric material that is electrically neutral and thus the center position of the gravity of positive charges at a volume element coincides with the center position of the gravity of negative electric charges in this volume (the first moment of electric charges is zero). In applied electric fields, positive and negative electric charges are displaced in opposite directions each other from their original positions (the first moment of electric charges becomes nonzero). In macroscopic electrostatics, dielectric polarizations are defined as vectors $\mathbf{P}(\mathbf{r})$, whose directions are defined to be in the direction of the displacement of electric charges and the magnitudes are (net) electric charges that pass through a unit area [Cm].

Bound electric charges Q_{pol} that are induced by dielectric polarizations $\mathbf{P}(\mathbf{r})$ have the form

$$\oint_{\Delta S} \mathrm{d}S \mathbf{P}(\mathbf{r}) \cdot \mathbf{n} = -Q_{\mathrm{pol}}, \tag{2.16}$$

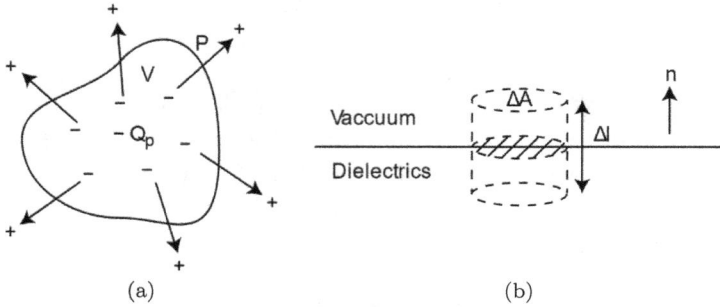

(a)   (b)

Fig. 2.1:   Dielectric polarizations $\mathbf{P}(\mathbf{r})$ represent displacements of the positions of positive electric charges relative to the positions of negative electric charges and thus leave negative charges $-Q_p$ in the volume V enclosed by a closed surface S (a). Gauss' pill box of cross section ΔA and height Δl is used to derive bound charges induced at the surfaces of dielectric materials (b). In the case of very small height Δl, the surface integrals of dielectric polarizations over the side of this pill box are zero.

in an arbitrary volume V that is enclosed by a closed surface S, where $\mathbf{n}$ is the normal unit vector of the surface (see Fig. 2.1(a)); dielectric polarizations leave negative charges behind.

Multiplying $1/\Delta V$ to both sides of Eq. (2.16) for a volume element ΔV, and taking the limit of $\Delta V \to 0$ leads to the differential form of Eq. (2.16);

$$\rho_{\mathrm{pol}}(\mathbf{r}) = -\nabla \cdot \mathbf{P}(\mathbf{r}), \qquad (2.17)$$

where Eq. (2.17) represents bound electric charges induced in the bulk of dielectric materials.

To derive the boundary conditions, we use the so-called Gauss' pill box, see Fig. 2.1(b), which is a cylindrical region of cross-section ΔA and height Δl across the surface. For the case of small height $\Delta l \to 0$; the surface integrals of dielectric polarizations over the side of this cylinder is zero and only bound surface charges q_{pol} at the surface of this dielectric material. By applying Eq. (2.16) to this region, the area density of bound surface charges is derived in the form

$$\sigma_{\mathrm{pol}}(\mathbf{r}) = \mathbf{P}(\mathbf{r}) \cdot \mathbf{n}(\mathbf{r}), \qquad (2.18)$$

where $\mathbf{n}$ is the normal vector of the surface and $\mathbf{r}$ is positional vectors to 2D surface (this notation is also used for positional vectors in the bulk (see Eq. (2.17))).

For the case of dielectric materials that show uniform dielectric polarizations, these dielectric polarizations induce bound charges at surfaces of these materials, Eq. (2.18), but not in the bulk, Eq. (2.17). We thus introduce dielectric polarizations as charges induced at the surfaces of dielectric materials. Gauss' divergence theorem ensures the charge neutrality condition that has the form

$$\int_V dV \rho_{\text{pol}}(\mathbf{r}) + \int_S dS \sigma_{\text{pol}}(\mathbf{r}) = 0. \tag{2.19}$$

We used a vector quantity $\mathbf{P}(\mathbf{r})$ to represent dielectric polarizations, because this is a mathematical expression representing the fact that one cannot extract bound charges from dielectric materials. Electric dipoles are the first moments of electric charges, see Eq. (2.13). Substituting Eqs. (2.18) and (2.19) into Eq. (2.13) leads to electric dipoles of a dielectric material in the form

$$\boldsymbol{\mu}_{\text{pol}} = \int_V dV \mathbf{P}(\mathbf{r}), \tag{2.20}$$

where Gauss' divergence theorem was used to derive the last equation. Dielectric polarizations are thus electric dipoles per unit volume.

Bound charges $\rho_{\text{pol}}(\mathbf{r})$ are induced in dielectric materials due to dielectric polarizations and thus reflect the properties of dielectric materials. In principle, it is possible to bring net electric charges $\rho_{\text{tru}}(\mathbf{r})$ into dielectric materials and one can discuss (macroscopic) electric fields that are generated by these electric charges or applied fields in dielectric materials. These electric charges $\rho_{\text{tru}}(\mathbf{r})$ that can be *in principle* extracted from or brought into dielectric materials are called *true* charges.

In dielectric materials, both *true* and *bound* charges, $\rho_{\text{tru}}(\mathbf{r})$ and $\rho_{\text{bou}}(\mathbf{r})$, respectively, generate electric fields and thus Eq. (2.1) is rewritten in the form

$$\nabla \cdot \mathbf{D}(\mathbf{r}) = \rho_{\text{tru}}(\mathbf{r}), \tag{2.21}$$

with

$$\mathbf{D}(\mathbf{r}) = \epsilon_0 \mathbf{E}(\mathbf{r}) + \mathbf{P}(\mathbf{r}). \tag{2.22}$$

Here Eq. (2.16) is used to derive Eq. (2.22). The vector fields $\mathbf{D}(\mathbf{r})$ are called electric displacement vectors or electric flux density (because the form of Eq. (2.21) implies that D is the number of "fluxes" generated from true charges $\rho_{tru}(\mathbf{r})$. Taking rotation to both sides of Eq. (2.14) and using the fact that microscopic electric fields are irrotational, see Eq. (2.2)), leads to the form

$$\nabla \times \mathbf{E}(\mathbf{r}) = \mathbf{0}. \tag{2.23}$$

Equations (2.21) and (2.23) consist of the fundamental laws of electrostatics in dielectric materials. In principle, one can calculate electric fields in dielectric materials once dielectric polarizations, $\mathbf{P}(\mathbf{r})$, are given as functions of electric fields by solving Eqs. (2.21) and (2.23) with boundary conditions (however, because of the fact that Eq. (2.21) is given as a function of electric displacement vectors and thus the fundamental theorem of vector calculus does not ensure that electric fields are determined uniquely). The form of dielectric polarizations $\mathbf{P}(\mathbf{r})$ as functions of (applied) electric fields depends on the properties of materials.

The boundary conditions for electric displacement vectors are derived by calculating the volume integrals of both sides of Eq. (2.21) with respect to a region enclosed by a cylindrical closed surface (so-called Gauss' pill box) of cross-section ΔA and height Δl that is across the boundary (see Fig. 2.2(a)). The height Δl of this closed surface is small, $\Delta l \to 0$; (bulk) true electric charges $\rho_{tru}\Delta A \delta l$ (that are given by the volume density ρ_{tru}) are not included in the pill box, but surface true charges $q_{tru}\Delta A$ at the boundary that are given by the area density q_{tru}) are in the pill box. We change the volume

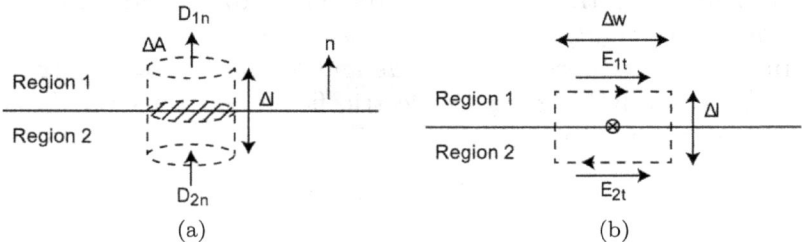

(a)

(b)

Fig. 2.2: Gauss' pill boxes to derive the boundary condition of electric displacement vectors (a) and electric fields (b) in the systems of dielectric materials. The volume integrals of Eq. (2.21) with respect to the region of pill box leads to the boundary conditions for the normal components of electric displacement vectors, Eq. (2.24).

integral of the divergence of electric displacement vectors to the surface integral of these vectors by using Gauss' divergence theorem and neglect electric displacement vectors that cross the sides of this pill box because $\Delta l \to 0$. This leads to the boundary conditions of electric displacement vectors in the form

$$q_{\text{tru}} = D_{1n} - D_{2n}, \qquad (2.24)$$

where D_{1n} ($\equiv \mathbf{D}_1 \cdot \mathbf{n}$) and D_{2n} ($\equiv \mathbf{D}_2 \cdot \mathbf{n}$) are the normal components of electric displacement vectors at region 1 and 2 sides of the boundary, respectively, and the normal $\mathbf{n}$ is directed from region 2 to 1, see Fig. 2.2(a). q_{tru} is the area density of (surface) true charges at the boundary. Equation (2.24) suggests that the normal components of electric displacement vectors are discontinuous across the boundary for the cases that there are surface true charges at the boundary. This reflects the fact that we treat electric fields that are averaged over a macroscopic length scale and thus neglect the thickness of interfaces.

The boundary conditions for electric fields are derived by calculating the surface integrals of both sides of Eq. (2.23) with respect to the rectangular area of width Δw and height Δl (see Fig. 2.2(b)). We change the surface integral of the rotations of electric fields to the line integrals of these fields by using Storks' theorem. This leads to the boundary conditions for the tangent component of electric fields

$$E_{1t} = E_{2t}, \qquad (2.25)$$

where E_{1t} and E_{2t} are the tangent components of electric fields; the tangent components of electric fields are continuous across the interfaces between dielectric materials.

In practice, dielectric polarizations induced in isotropic 3D bulk materials are approximately linear with respect to the applied electric fields. This behavior can be phenomenologically described by the following equation:

$$\mathbf{P}(\mathbf{r}) = \epsilon_0 \chi^{(1)} \mathbf{E}(\mathbf{r}), \qquad (2.26)$$

where the coefficient of proportionality $\chi^{(1)}$ is called linear susceptibility. Substituting Eq. (2.26) into Eq. (2.22) leads to the form

$$\mathbf{D}(\mathbf{r}) = \epsilon \mathbf{E}(\mathbf{r}). \qquad (2.27)$$

Here ϵ ($= \epsilon_0(1 + \chi^{(1)})$) is dielectric constant. In general, dielectric constant ϵ is a function of positions and time. For the case of uniform dielectric materials, Eq. (2.21) gives the divergence of electric fields and thus the fundamental theorem of vector calculus ensures that Eqs. (2.21) and (2.23) uniquely determine electric fields with boundary conditions. As macroscopic electric fields are irrotational, Eq. (2.23), macroscopic electric fields are represented by the negative of the gradient of (macroscopic) electrostatic potentials:

$$\mathbf{E}(\mathbf{r}) = -\nabla\phi(\mathbf{r}), \qquad (2.28)$$

where macroscopic electrostatic potentials, $\phi(\mathbf{r})$, are defined in Eq. (2.15). Equations (2.21), (2.27), and (2.28) lead to Poisson equation in dielectric materials:

$$\nabla^2\phi(\mathbf{r}) = -\frac{\rho_{\text{tru}}(\mathbf{r})}{\epsilon}. \qquad (2.29)$$

Indeed, the form of Eq. (2.29) resembles Eq. (2.4); when solutions to Eq. (2.5) are known for an electric charge distribution and a boundary condition, solutions to Eq. (2.29) for the same true charge distribution and boundary condition can be derived by simply replacing ϵ_0 with ϵ. The effects of dielectric polarizations are encompassed in the dielectric constant ϵ.

2.1.3.   *Electrostatic energy and Maxwell stress tensor*

Electrostatic energy can be easily derived by analyzing parallel capacitors composed of ideal isotropic dielectric material, see Appendix A. The derivation implies that electrostatic energy is generated because of the presence of electric charges induced on the electrodes, meanwhile it also implies that electrostatic energy is stored in dielectric material subjected to external electric fields.

We here treat electrostatic energy in dielectric material more generally. This electrostatic energy is composed of two parts; electrostatic energy W_{ele} arising from the fact that electric fields are generated in dielectric material and work done W_{pol} to generate dielectric polarizations. The electrostatic energy W_{ele} has the form

$$W_{\text{ele}} = \frac{1}{2}\int_s dS\sigma(\mathbf{r})\phi(\mathbf{r}) + \frac{1}{2}\int_V dV\rho(\mathbf{r})\phi(\mathbf{r}), \qquad (2.30)$$

where $\sigma(\mathbf{r})$ is the area density of surface electric charges at the boundary and $\rho(\mathbf{r})$ is the volume density of electric charges in the bulk of the system. $\sigma(\mathbf{r})$ and $\rho(\mathbf{r})$ are total electric charges that include both true and bound charges. Equation (2.30) represents electrostatic energy arising from electrostatic Coulomb interactions in the limit of continuum description and the factor of $1/2$ in the both terms is to avoid double counting of these interactions; electrostatic potentials $\phi(\mathbf{r})$ are generated by electric charges $\sigma(\mathbf{r})$ and $\rho(\mathbf{r})$. We use Gauss' law, Eq. (2.1), to the second term of Eq. (2.30) and Gauss' divergence theorem to change the volume integral of $\nabla \cdot (\phi(\mathbf{r})\boldsymbol{E}(\mathbf{r}))$ to the surface integral. Because $\boldsymbol{E}(\mathbf{r}) \cdot \mathbf{n}(\mathbf{r})$ is $-\sigma(\mathbf{r})/\epsilon_0$ (where $\mathbf{n}(\mathbf{r})$ is the outward normal of the boundary), Eq. (2.30) is rewritten in the form

$$W_{\text{ele}} = \frac{1}{2}\epsilon_0 \int_V dV E^2(\mathbf{r}), \tag{2.31}$$

where $E(r)$ $(\equiv |\boldsymbol{E}(\mathbf{r})|)$ is the magnitudes of electric fields.

The work that applied electric fields $\mathbf{E}(\mathbf{r})$ perform by displacing a positive charge relative to a negative charge by dl is $q\mathbf{E}(\mathbf{r}) \cdot dl$. Dielectric polarizations are electric dipoles per unit volume, $d\mathbf{P}(\mathbf{r}) = qN_m dl$, and the work w_{pol} per unit volume done by generating dielectric polarizations thus has the form (see Eq. (2.20))

$$dW_{\text{pol}(\mathbf{r})} = \boldsymbol{E}(\mathbf{r}) \cdot d\mathbf{P}(\mathbf{r}). \tag{2.32}$$

We here did not consider the term $-\mathbf{E}(\mathbf{r}) \cdot \mathbf{P}(\mathbf{r})dN_m/N_m$; the volume density N_m of molecules does not change during the process of generating dielectric polarizations. This term must be considered for the case of dielectric materials, where dielectric polarizations depend on the volume density of molecules (and/or their crystal structures for the case of crystals) (see Landau and Lifshitz [5]). Except for the latter cases, Eq. (2.32) is quite general. To continue the calculations, it is necessary to specify the functional form of dielectric polarizations $\mathbf{P}(\mathbf{r})$. For the case of isotropic 3D bulk materials, $S_1 = 0$, $S_2 = 0$, and $S_3 = 0$, that only generate linear polarizations, and the work done by generating dielectric polarizations has the form

$$W_{\text{pol}} = \frac{1}{2} \int_V dV \mathbf{E}(\mathbf{r}) \cdot \mathbf{P}(\mathbf{r}), \tag{2.33}$$

where we substituted Eq. (2.26) into Eq. (2.32) to derive Eq. (2.33).

The sum of Eqs. (2.31) and (2.33) is equal to the total electrostatic energy with the volume density of electrostatic energy $\frac{1}{2}\epsilon E^2$, and it can be shown that tensional Maxwell stresses of $\frac{1}{2}\epsilon E^2$ are generated in the direction parallel to electric fields and compressional Maxwell stresses of $-\frac{1}{2}\epsilon E^2$ are generated in the direction perpendicular to electric fields (see Appendix A). These results are general, and not limited to the case with the geometry of parallel plate capacitors.

The derived Maxwell stresses are summarized in Maxwell stress tensor, whose (i, j) component has the form

$$T_{ij} = \epsilon \left(E_i(\mathbf{r})E_j(\mathbf{r}) - \frac{1}{2}E^2(\mathbf{r})\delta_{ij} \right), \tag{2.34}$$

where i and j are x, y, and z (the (x, y) components are $i = x$ and $j = y$), $E = (E_x, E_y, E_z)$ are the components of electric fields, and δ_{ij} is Kronecker's delta (1 for $i = j$ and 0 for $i = j$). Maxwell stresses are forces per unit area applied to the surfaces of volume elements in the dielectric materials, and thus net forces applied to volume elements have the form

$$\mathbf{f}_V(\mathbf{r})\Delta V \equiv \mathbf{e}_i \frac{\partial}{\partial x_j}T_{ij}\Delta V = \rho_{\text{tru}}(\mathbf{r})\Delta V \, \mathbf{E}(\mathbf{r}), \tag{2.35}$$

with i and j are x, y, and z, where $\mathbf{e}_x$, $\mathbf{e}_y$, and $\mathbf{e}_z$ are the unit vectors to the x-, y-, and z-directions. Here and after, we use the Einstein summation convention, where a pair of indices of the same character should be summed over x, y, and z.

Equation (2.35) implies that electrostatic Maxwell stresses do not generate net electric forces in the bulk of dielectric materials unless there are *true charges* in the bulk. The area density of net electric forces applied to the surfaces of dielectric materials have the form

$$\mathbf{f}_A(\mathbf{r}) = (f_{Ax}, f_{Ay}, f_{Az}) \quad \text{with } f_{Ai} = T_{ij}n_j, \tag{2.36}$$

for $i = x$, y, and z, where $\mathbf{n} = (n_x, n_y, n_z)$ is the outward normal of the surfaces of this dielectric material.

Equation (2.36) leads to the area density of tangential (shear) forces applied to interfaces between two dielectric materials thus have the form

$$\mathbf{f}_{\parallel} = q_{\text{tru}}(\mathbf{r})\mathbf{E}_{\parallel}, \tag{2.37}$$

with components of electric fields. Equation (2.37) implies that electrostatic Maxwell stresses do not generate net tangential forces unless there are true charges at the interfaces between dielectric materials. Noteworthy that many textbooks treat the cases in which true surface charges are not accumulated at interfaces between dielectric materials, $q_{tru}(\mathbf{r}) = 0$, and thus conclude that electrostatic Maxwell stresses *do not* generate net shear electric forces (see Refs. [1–3]). Indeed, true charges are accumulated at interfaces between lossy dielectric materials due to interfacial polarizations (or Maxwell–Wagner eflect) [6]. This implies that Maxwell stresses apply net shear electric forces to interfaces between lossy dielectric soft materials and contribute to deform the shapes of these materials. It is thus of interest to study the electromechanical properties of lossy dielectric soft materials in terms of the shape deformation, e.g., vesicles in solution.

2.2. Microscopic Mechanisms of Dielectric Polarizations in Isotropic Materials

2.2.1. *Linear polarizability and local electric fields*

Applied electric fields induce electric dipoles and this is the microscopic mechanism of dielectric polarizations, see Appendix A. There are four typical mechanisms: electronic polarizations, ionic polarizations (atomic polarizations), orientational polarizations (dipolar polarizations), and interfacial polarizations [7–9], but we here focus on electronic polarizations and dipolar polarizations, because these polarizations are particularly important to analyze domains in organic monolayers, experimentally and theoretically.

Isotropic materials show linear polarizations (see Eq. (2.26)). This means that average electric dipoles generated in constituent molecules are linear to applied electric fields, and an electric dipole $\boldsymbol{\mu}_{lin}$ induced in a constituent molecule is written in the form

$$\boldsymbol{\mu}_{lin} = \alpha_{iso}\boldsymbol{E}_{loc}. \tag{2.38}$$

Here E_{loc} is local electric field, and different from macroscopic electric fields $E(r)$. Indeed, the local electric field is given by $\boldsymbol{E}_{loc} = \boldsymbol{E} + \frac{\boldsymbol{P}}{3\epsilon_0}$ for the case of isotropic dielectric materials [8, 9]. This local field

E_{loc} is called Lorentz molecular electric field, where the second term of the field $P/3\epsilon_0$ appears self-consistently due to the polarization of constituent molecules and atoms in isotropic materials. The coefficient of proportionality α iso is called (linear) polarizability and reflects the microscopic dynamics of electrons and nuclei in constituent molecules.

2.2.2. *Electronic polarizability*

Electronic polarizations are one of the mechanisms of dielectric polarizations, where local electric fields induce electric dipoles by displacing the center of gravity of electrons of molecules (that have negative charges) from the position of nuclei (that have positive charges). The knowledge of quantum chemistry is necessary to quantitatively depict the linear polarizability. However, a simple classical model is sufficient to capture the physics of electronic polarizations.

Let us assume spherical atoms, where their nuclei are at their center and electrons are uniformly distributed in the spherical region of radius a_0 (see Fig. 2.3). It is convenient to use a coordinate system, whose origin is located at (and thus moves together with) the center of gravity of electrons (see Fig. 2.3). In the absence of local electric fields $\mathbf{E}_i$, the center of gravity of electrons is at the position of nucleus. In applied local electric fields, the center of gravity of electrons is displaced from the position of nuclei until electric

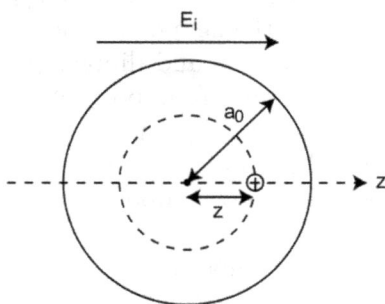

Fig. 2.3: A simple classical model of electronic polarizations treats spherical atoms of radius a_0. Local electric fields displace the centers of the gravity of electrons (that have negative charges) from the position of nucleus (that has positive charges) by z. We use a coordinate system whose origin is at the center of gravity of the electrons.

forces arising from local electric fields $\mathbf{E}_i$ and electric forces arising from interactions between electrons and atoms are balanced. The magnitude of electric fields applied to the nucleus at $r = z$ has the form [8, 9]

$$E_r = -\frac{Q_{\text{eff}}}{4\pi\epsilon_0 z^2},\qquad(2.39)$$

with

$$Q_{\text{eff}} = Q\frac{\frac{4\pi}{3}z^3}{\frac{4\pi}{3}a_0^3} = Q\frac{z^3}{a_0^3},\qquad(2.40)$$

by assuming that only electric charges of electrons at $r < z$ contributes to the electric fields at $r = z$. The balance of forces applied to the nucleus (and thus the center of the gravity of electrons) has the form

$$QE_i - \frac{Q^2}{4\pi\epsilon_0 a_0^3}z = 0.\qquad(2.41)$$

The displacement z is derived by solving Eq. (2.41); the displacement z multiplied by the electric charge Q of the nucleus is the dipole moment μ_{ele} that is induced by local electric fields $\mathbf{E}_i$. Using Eq. (2.38), (linear) electronic polarizability α_{ele} is derived:

$$\alpha_{\text{ele}} = 4\pi\epsilon_0 a_0^3.\qquad(2.42)$$

Electronic polarizability is proportional to the volume of individual atoms (and molecules).

2.2.3. *Linear orientational polarizability [7–9]*

Orientational polarizations are generated due to the competition between thermal energy and electrostatic interactions. We here show the free rotational model of orientational polarizations, where interactions between constituent molecules, including long-range electrostatic dipole–dipole interactions, are discarded (see Fig. 2.4). It is convenient to use a coordinate system, where the z-direction is parallel to the direction of local electric fields. The electrostatic energy

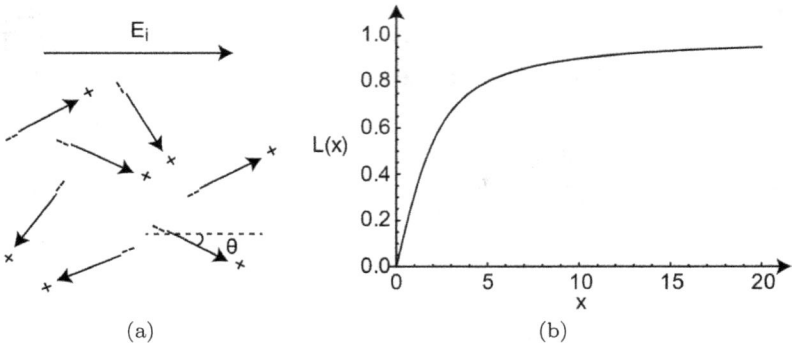

Fig. 2.4: An orientational gas model of orientational polarizations (a). It is convenient to use a coordinate system, where the z-direction is parallel to local electric fields. The orientations of permanent dipoles are specified by polar angles θ between individual permanent dipoles and the z-direction. Langevin function $L(x) = \coth x - 1/x$ (b).

of individual molecules due to interactions between their permanent dipoles $\boldsymbol{\mu}$ and local electric fields $\mathbf{E}_i$ have the form (see Eq. (2.11)):

$$W_{\text{ori}}(\theta, \phi) = -\mu \mathbf{E}_i \cos \theta, \qquad (2.43)$$

where θ is polar angle between individual permanent dipoles and local electric fields. For the cases of dilute gases, where the permanent dipoles of constituent gas molecules can orient in all directions in 3D spaces, the probability of finding polar molecules, where their permanent dipoles have an angle θ with local electric fields $\mathbf{E}_i$, is given by Boltzmann distribution that has the form

$$f(\theta, \phi) = \frac{1}{Z} e^{-\beta W_{\text{ori}}(\theta, \phi)}, \qquad (2.44)$$

with

$$Z = \int d\phi \int d(\cos \theta) e^{-\beta W_{\text{ori}}(\theta, \phi)}, \qquad (2.45)$$

where, Z is the partition function and $\beta \ (\equiv 1/k_B T)$ is the inverse of thermal energy. Here $k_B \ (= 1.38 \times 10^{-23} \ [\text{J/K}])$ is the Boltzmann constant and T is the absolute temperature. Many thermodynamic quantities are calculated using Eq. (2.44). The thermodynamic average of electric dipoles $\boldsymbol{\mu} = (\mu_x, \mu_y, \mu_z)$ has the forms

$$\mu_x = \mu_y = 0,$$

and

$$\mu_z = \mu\langle\cos\theta\rangle = \mu L(\beta\mu E_i), \tag{2.46}$$

where the function

$$L(x) = \cot hx - \frac{1}{x}, \tag{2.47}$$

is called the Langevin function. See Fig. 2.4(b) where $\coth x = 1/\tanh x$. The average electric dipole is proportional to the orientational order of permanent dipoles in the direction to the local electric fields $\mathbf{E}_i$, see the second term of Eq. (2.46). Equation (2.46) shows that $\mu E_i/k_B T$ is an important parameter that determines the z-component of electric dipoles $\langle\mu_z\rangle$, where μE_i is (maximum) electrostatic energy arising from the interactions between local electric fields and permanent electric dipoles and $k_B T$ is thermal energy; the competition between them is the physical mechanism involved in orientational polarizations. In general experimental conditions, the electrostatic energy arising from interactions between local electric fields and permanent dipoles is much smaller than thermal energy, $\mu_0 E_i/k_B T \ll 1$; the thermal energy at room temperature is in the order of 4.1×10^{-21} J ($= 26$ meV), whereas the electrostatic energy W_{ori} is typically in the order of 3.3×10^{-24} J for the cases that permanent dipoles of molecules are 1D and the magnitudes of local electric fields are 1.0×10^6 V/m. In these cases of $\mu_0 E_i/k_B T \ll 1$, Eq. (2.46) has an asymptotic form

$$\langle\mu_z\rangle \simeq \alpha_{\text{ori}} E_i, \tag{2.48}$$

with

$$\alpha_{\text{ori}} = \frac{\mu_0^2}{3k_B T}. \tag{2.49}$$

We used an asymptotic form $L(x) \cong \frac{1}{3}x$ of the Langevin function for $x \ll 1$. α_{ori} is linear orientational polarizability arising from orientational polarizations. The 3 in denominator reflects the fact that permanent dipoles can orient in any direction in 3D spaces.

2.3.   Dielectric Polarizations of Monolayers with C_∞-Symmetry [9, 13]

2.3.1.   *Dielectric polarizations of solids*

The general form of dielectric polarizations of solids is given by [10–12]

$$\mathbf{P} = \mathbf{P_0} + \epsilon_0 \chi^{(1)} \cdot \mathbf{E} + \epsilon_0 \chi^{(2)} : \mathbf{EE} + \cdots , \qquad (2.50)$$

where this is the simple Taylor expansion of dielectric polarizations $\mathbf{P}$ with respect to applied electric fields $\mathbf{E}$.

The first term $\mathbf{P_0}$ is called spontaneous polarizations that are generated in the absence of applied electric fields ($\mathbf{E} = \mathbf{0}$), e.g., monolayers on water surface have $\mathbf{P_0}$ ($\neq 0$).

The second term $\mathbf{P}^{(1)} (\equiv \epsilon_0 \chi^{(1)} \cdot \mathbf{E})$ is linear polarizations that are generated in proportion to applied electric fields $\mathbf{E}$. This term corresponds to electronic polarizations and dipolar orientational polarizations that are in proportion to applied electric fields (see 2.1.1). Generally, the coefficient $\chi^{(1)}$ of proportionality, linear susceptibility, are represented by second-rank tensors with consideration of the anisotropy of the system; this is in contrast to the case of isotropic materials, where linear susceptibility is given by a scaler quantity. The inner product $\chi^{(1)} \cdot \mathbf{E}$ is defined as $[\chi^{(1)} \cdot \mathbf{E}]_i \equiv \chi_{ij}^{(1)} E_j$, where i and j are x, y, and z, and $[\mathbf{A}]_i$ is the ith component of arbitrary vector $\mathbf{A}$.

The third term $\mathbf{P}^{(2)} (\equiv \epsilon_0 \chi^{(2)} : \mathbf{EE})$ is called second-order nonlinear polarizations that are generated in proportion to the square of applied electric fields. The coefficients of proportionality $\chi^{(2)}$ are third-rank tensors called second-order nonlinear susceptibility (SOS). The inner product $\chi^{(2)} : \mathbf{EE}$ is defined as $[\chi^{(2)} : \mathbf{EE}]_i = \chi_{ijk}^{(2)} E_j E_k$, where i, j, and k are x, y, and z. Indeed, it is possible to show that spontaneous polarizations and second-order nonlinear polarizations are not generated from centrosymmetric bulk materials, e.g., isotropic liquids and nematic liquid crystals, meanwhile they are generated from non-centrosymmetric materials such as monolayers.

For centrosymmetric materials cases, they do not change their properties by the inverse operation $x \to -x$, $y \to -y$, and $z \to -z$. When applied electric fields are reversed $E \to -E$, dielectric polarizations are generated in the opposite direction:

$$-\mathbf{P} = \mathbf{P}_0 - \epsilon_0 \chi^{(1)} \cdot \mathbf{E} + \epsilon_0 \chi^{(2)} : \mathbf{EE} + \cdots . \tag{2.51}$$

Equations (2.50) and (2.51) are satisfied simultaneously only when both $\mathbf{P}_0$ and $\chi^{(2)} : \mathbf{EE}$ are zero. That is, $\chi^{(2)} = 0$.

2.3.2. Dielectric polarizations of monolayers composed of rod-shaped molecules

Spontaneous polarizations $\mathbf{P}_0$, linear susceptibilities $\chi^{(1)}$, and second-order nonlinear susceptibilities $\chi^{(2)}$ depend on the electronic properties of constituent molecules and the orientational orders. A simple microscopic model of monolayers composed of rod-shaped molecules is convenient to derive explicit forms of these dielectric polarizations as functions of orientational orders, i.e., S_1, S_2, and S_3. This model can apply to a variety of monolayers such as fatty acids, phospholipids, and cyanobiphenyls, etc., where their hydrophilic heads and hydrophobic chains are linearly bonded. Note that the cases of molecules having more complex shapes, e.g., banana-shaped molecules, must be treated as an extension of the case of rod-shaped molecules.

Owing to thermal free rotations of the constituent rod-shaped molecules around their long-axes, the molecules are treated to have cylindrical symmetry around their long-axes. On the other hand, the structures of these molecules are not symmetric with respect to the inversion of their long-axes; constituent molecules themselves have C_∞-symmetry. These polar molecules thus have permanent electric dipoles and second-order hyperpolarizability.

In applied electric fields, constituent molecules in monolayers induce electric dipoles that have the form

$$\boldsymbol{\mu}_{\text{tot}} = \boldsymbol{\mu}_0 + \alpha^{(1)} \cdot \mathbf{E}_{\text{loc}} + \beta^{(2)} : \mathbf{E}_{\text{loc}} \mathbf{E}_{\text{loc}} + \cdots . \tag{2.52}$$

The first term $\boldsymbol{\mu}_0$ is permanent dipoles. The second term is electric dipoles that are linear to local electric fields $\mathbf{E}_{\text{loc}}$. The coefficients $\alpha^{(1)}$ of proportionality are second-rank tensors called linear polarizabilities and these linear polarizabilities only include the contributions of electronic polarizations. The third term is proportional to the square of local electric fields. The coefficients $\beta^{(2)}$ of proportionality are second-order hyperpolarizabilities that are represented by third-rank tensors and these second-order hyperpolarizabilities only include the contributions of electronic polarizations.

2.3.2.1. *Physical quantities of rod-shaped molecules*

The physical quantities of rod-shaped molecules have simple forms when they are represented in the molecular coordinate system, where the w-direction is parallel to the long-axis of the molecule and the u- and v-directions are perpendicular to the long-axis of the molecule (see Fig. 2.5). These are the principle axes of a rod-shaped molecule and thus is useful to represent the properties of individual molecules.

(a-1) Permanent dipoles:

Because of the cylindrical symmetry of rod-shaped constituent molecules, the permanent dipole moment is directed in w-direction and thus has the form

$$\boldsymbol{\mu} = \mu_0(0, 0, 1)^m, \qquad (2.53)$$

where μ_0 is the moment of this permanent dipole and $\mathbf{e}_w$ is the unit vector to w-direction. The superscript m stands for the molecular coordinate (see Fig. 2.5). Note that, μ_0 represents the w-component

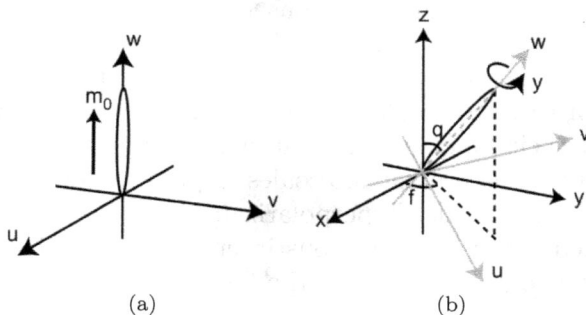

(a)   (b)

Fig. 2.5:   A molecular model to analyze the relationship between dielectric polarizations generated in monolayers and the orientational order parameters S_1, S_2, and S_3. We treat the case of monolayers that are composed of polar rod-shaped molecules that have cylindrical symmetry. We define the molecular coordinate system, where the w-axis is parallel to the long-axis of a molecule and the u- and v-axes are parallel to its short-axes (a) and the monolayer coordinate system, where z-axis is parallel to the monolayer normal and the x- and y-axes are in the monolayer plane (b). These coordinate systems are transformed from one to the other by using the Euler matrix.

of the permanent dipole of a constituent molecule because the dipole moment in the u- and v-axes are eliminated owing to the thermal free rotations of this molecule around its long-axis.

(a-2) Linear polarizability:

Linear polarizability $\alpha^{(1)}$ has a matrix form

$$
\alpha = \begin{pmatrix} \alpha_\perp & \alpha_{ch} & 0 \\ -\alpha_{ch} & \alpha_\perp & 0 \\ 0 & 0 & \alpha_\| \end{pmatrix}^m , \tag{2.54}
$$

where $\alpha_\perp$ and $\alpha_\|$ are the components of linear polarizability in the short- and long-axes of the molecule, respectively, and α_{ch} is the "chiral" component of linear polarizability that is absent from achiral molecules; $\alpha_{uu} = \alpha_{vv} = \alpha_\perp$, $\alpha_{ww} = \alpha_\|$, $\alpha_{uv} = -\alpha_{vu} = \alpha_{ch}$, and the other components are zero.

(a-3) Second-order hyperpolarizability:

The second-order hyperpolarizability $\beta^{(2)}$ has the matrix form

$$
\beta^{(2)} = \begin{pmatrix} \beta_{11} & \beta_{12} & \beta_{13} & \beta_{14} & \beta_{15} & \beta_{16} \\ \beta_{21} & \beta_{22} & \beta_{23} & \beta_{24} & \beta_{25} & \beta_{26} \\ \beta_{31} & \beta_{32} & \beta_{33} & \beta_{34} & \beta_{35} & \beta_{36} \end{pmatrix}^m
$$

$$
= \begin{pmatrix} 0 & 0 & 0 & \beta_{14} & \beta_{15} & 0 \\ 0 & 0 & 0 & \beta_{15} & -\beta_{14} & 0 \\ \beta_{31} & \beta_{31} & \beta_{33} & 0 & 0 & 0 \end{pmatrix}^m , \tag{2.55}
$$

where $\beta_{i1} = \beta_{i11}^{(2)}$, $\beta_{i2} = \beta_{i23}^{(2)}$, $\beta_{i3} = \beta_{i33}^{(2)}$, $\beta_{i4} = (\beta_{i23}^{(2)} + \beta_{i32}^{(2)})/2$, $\beta_{i5} = (\beta_{i31}^{(2)} + \beta_{i13}^{(2)})/2$, and $\beta_{i6} = (\beta_{i12}^{(2)} + \beta_{i21}^{(2)})/2$ for $i = 1$, 2, and 3. The subscripts 1, 2, and 3 here stand for u, v, and w, respectively; we use this rule for vector and tensor quantities that are defined in the molecular coordinate system. β_{14} is the chiral component that is absent from achiral molecules. The forms of Eqs. (2.53), (2.54), and (2.55) are derived by using the fact that constituent molecules have C_∞-symmetry.

The monolayer coordinate system, where the z-direction is parallel to the monolayer normal and the x- and y-directions are parallel to the monolayer plane, is useful to treat the physical quantities of monolayers that are assemblies of molecules. We use the Euler rotational matrix that has the form

$$R(\theta, \psi, \phi)$$
$$= \begin{pmatrix} \cos\phi\cos\theta\cos\psi - \sin\phi\sin\psi & -\sin\psi\cos\phi\cos\theta - \sin\phi\cos\psi & \cos\phi\sin\theta \\ \sin\phi\cos\theta\cos\psi + \cos\phi\sin\psi & -\sin\psi\sin\phi\cos\theta + \cos\psi\cos\phi & \sin\phi\sin\theta \\ -\sin\theta\cos\psi & \sin\theta\sin\psi & \cos\theta \end{pmatrix},$$

$$(2.56)$$

to represent the dipoles in the monolayer coordinate system (x, y, z), where $\theta, \psi,$ and ϕ are tilt angle, twist angle, and tilt azimuth, respectively. The transformation by the Euler matrix is composed of three independent rotation; the molecular coordinate is first rotated around the w-axis by angle ψ to a new coordinate system $(x'', y'', z'')^{\text{fir}}$, this new coordinate system $(x'', y'', z'')^{\text{fir}}$ is rotated around y''-axis by angle θ to another coordinate system $(x', y', z')^{\text{seco}}$, and finally, the last coordinate system $(x', y', z')^{\text{seco}}$ is rotated around the z'-axis by angle ϕ (see Fig. 2.5(b)).

In summary, the permanent dipole, linear polarizability, and second-order nonlinear hyperpolarizability of a constituent molecule are represented in the monolayer coordinate system in the form

$$\mu_i = \sum_{l=u,v,w} R_{il}(\theta, 0, \phi)\mu_l, \qquad (2.57)$$

$$\alpha_{ij} = \sum_{l,m=u,v,w} R_{il}(\theta, 0, \phi)R_{jm}(\theta, 0, \phi)\alpha_{lm}, \qquad (2.58)$$

$$\beta_{ijk} = \sum_{l,m,n=u,v,w} R_{il}(\theta, 0, \phi)R_{jm}(\theta, 0, \phi)R_{kn}(\theta, 0, \phi)\beta_{lmn}, \qquad (2.59)$$

where i, j, and k are x, y, and z.

2.3.2.2. *Dielectric polarizations of monolayers*

Dielectric polarizations of monolayers are the vector sum of the electric dipoles of constituent molecules in a unit area. Monolayers are composed of an enormous number of molecules, typically in the order of 10^{14}–10^{15} molecules in the square of 1 cm $\times$ 1 cm, and thus we use the statistical mechanics to derive the form of dielectric polarizations.

The orientational distribution function of a monolayer that has C_∞-symmetry (and their constituent molecules thus do not show *in-plane* orientational ordering) has an expanded form

$$f(\theta,\phi) = \frac{1}{2\pi} \sum_{k=0}^{\infty} \frac{2k+1}{2} S_k P_k(\cos\theta), \qquad (2.60)$$

where $P_n(\cos\theta)$ is the Legendre polynomials of nth order and S_n is the orientational order parameter of nth order (see also Section 1.2).

(b-1) Spontaneous polarization [10, 13]:

Dielectric polarizations generated from monolayers have the form [10, 13]

$$\mathbf{P_0} = N_s \langle \mathbf{m} \rangle, \qquad (2.61)$$

where N_s is the number of molecules in a unit area and $\langle\ \rangle$ is the thermodynamic average with respect to the orientational distribution function (Eq. (2.60)).

For monolayers with the C_∞-symmetry, where these monolayers are symmetric with respect to rotations around their normal by any degrees, but are not symmetric with respect to the inversion of the space, spontaneous polarizations have the form

$$\mathbf{P_0} = N_s \mu_0 S_1 \mathbf{n}, \qquad (2.62)$$

where $\mathbf{n}$ is the unit vector that is normal to the monolayers. Equation (2.62) is derived by substituting Eqs. (2.57) and (2.60) into Eq. (2.61). Because spontaneous polarizations appear even in the absence of applied electric fields, the orientational order parameter S_1 in Eq. (2.62) is the value for the case of $\mathbf{E}_{\text{loc}} \to 0$ in Eq. (2.62); the higher-order powers of S_1 with respect to local electric fields $\mathbf{E}_{\text{loc}}$ contribute to higher-order polarizations (orientational polarizations). Spontaneous polarizations are proportional to the orientational order parameter S_1 and thus are not generated from centrosymmetric bulk materials, $S_1 = 0$, e.g., isotropic liquids and nematic liquid crystals. Monolayers of C_∞-symmetry generate spontaneous polarizations in the monolayer normal. This implies that monolayers store electrostatic energies that are necessary to generate these spontaneous polarizations.

(b-2) Linear electronic polarization [9–13]:

Dielectric polarizations that are linear to applied electric fields have the form (see the second term of Eq. (2.36)),

$$\mathbf{P}^{(1)} = \chi^{(1)} \cdot \mathbf{E}. \tag{2.63}$$

Although, in general, local electric fields applied to constituent molecules are different from applied electric fields, for simplicity, we treat the case of dilute monolayers, where electric fields arising from the dipoles of molecules at the neighbor are negligible to a good approximation, $\mathbf{E_{loc}} \cong \mathbf{E}$. Electronic polarizations are generated by the second term of Eq. (2.52) and these polarizations thus have the form

$$\begin{pmatrix} P_x^{(1)} \\ P_y^{(1)} \\ P_z^{(1)} \end{pmatrix} = \begin{pmatrix} \chi_\parallel^{(1)} & \chi_{ch}^{(1)} & 0 \\ \chi_{ch}^{(1)} & \chi_\parallel^{(1)} & 0 \\ 0 & 0 & \chi_\perp^{(1)} \end{pmatrix} \begin{pmatrix} E_x \\ E_y \\ E_z \end{pmatrix}, \tag{2.64}$$

in the monolayer coordinate system (x, y, z), where $\chi_\parallel^{(1)} = \chi_{xx}^{(1)} = \chi_{yy}^{(1)}$, $\chi_\perp^{(1)} = \chi_{zz}^{(1)}$ and $\chi_{ch}^{(1)} = \chi_{xy}^{(1)} = -\chi_{yx}^{(1)}$. The components $\chi_\parallel^{(1)}$, $\chi_\perp^{(1)}$, and $\chi_{ch}^{(1)}$ have the forms

$$\chi_\perp^{(1)} = N_s \left(\frac{2}{3}(\alpha_\parallel - \alpha_\perp)S_2 + \frac{2\alpha_\perp + \alpha_\parallel}{3} \right), \tag{2.65}$$

$$\chi_\parallel^{(1)} = N_s \left(-\frac{\alpha_\parallel - \alpha_\perp}{3}S_2 + \frac{2\alpha_\perp + \alpha_\parallel}{3} \right), \tag{2.66}$$

$$\chi_{ch}^{(1)} = N_s S_1 \alpha_{ch}. \tag{2.67}$$

These equations are derived by substituting Eq. (2.58) (that is the second term of Eq. (2.52)) into Eq. (2.61), where the thermodynamic average is calculated by using the orientation distribution function given by Eq. (2.60). The components $\chi_\parallel^{(1)}$ and $\chi_\perp^{(1)}$ are relevant to linear polarizations that are parallel and perpendicular to monolayers, respectively. $\chi_\parallel^{(1)}$ and $\chi_\perp^{(1)}$ are achiral components that are non-zero for both monolayers of achiral molecules and monolayers of chiral molecules. The difference $\chi_{\parallel}^{(1)} - \chi_{\perp}^{(1)} = N_s(\alpha_{\parallel} - \alpha_{\perp})S_2$ is proportional to an orientational order parameter (nematic order parameter) S_2 and the average $(2\chi_{\perp}^{(1)} + \chi_{\parallel}^{(1)})/3 = N_s(2\alpha_{\perp} + \alpha_{\parallel})/3$ is independent of this order parameter S_2 (see Eqs. (2.66) and (2.65)). $\chi_{ch}^{(1)}$ is chiral component and is zero for the cases of monolayers composed of achiral molecules. This component is proportional to an orientational order parameter S_1 and thus is zero for the cases of bulk materials that are centrosymmetric. Substituting Eq. (2.64) into Eq. (2.63) leads to the vector form of linear polarizations

$$\mathbf{P}^{(1)} = \chi_{\parallel}^{(1)}\mathbf{E} + (\chi_{\perp}^{(1)} - \chi_{\parallel}^{(1)})(\mathbf{E} \cdot \mathbf{n})\mathbf{n} + \chi_{ch}^{(1)}\mathbf{E} \times \mathbf{n}. \qquad (2.68)$$

For the case that orientational order parameters S_1 and S_2 are zero, Eq. (2.68) returns to linear polarizations of isotropic materials, see Eq. (2.26). For the case that an orientational order parameter S_1 is zero, Eq. (2.68) returns to the form of linear polarizations generated from nematic liquid crystals, where, in this case, $\mathbf{n}$ is director (the average direction of the long-axes of rod-shaped molecules). Monolayers and nematic liquid crystals show anisotropic linear polarizations because their orientational order parameter S_2 is not zero.

(b-3) Second-order nonlinear polarization [10–13]:

Dielectric polarizations that are proportional to the square of applied electric fields have the form (see the third term of Eq. (2.50))

$$\mathbf{P}^{(2)} = \chi^{(2)} : \mathbf{EE}. \qquad (2.69)$$

We here treat the case of dilute monolayers, where electric fields arising from the dipoles of constituent molecules are negligible to a good approximation. In general, there are contributions from the first, second, and third terms of Eq. (2.52). In many cases, second-order nonlinear polarizations are relatively small and high-power lasers are used to generate these polarizations. In this case, orientational polarizations are very small because of dielectric dispersion. We thus treat second-order nonlinear polarizations that originate from electronic polarizations, the third term of Eq. (2.52). The second-order nonlinear polarizations $\mathbf{P}^{(2)} = (P_x^{(2)}, P_y^{(2)}, P_z^{(2)})$ of monolayers that show

C_∞-symmetry have the form

$$
\begin{pmatrix} P_x^{(2)} \\ P_y^{(2)} \\ P_z^{(2)} \end{pmatrix} = \begin{pmatrix} 0 & 0 & 0 & s_{14} & s_{15} & 0 \\ 0 & 0 & 0 & s_{15} & -s_{14} & 0 \\ s_{33} & s_{31} & s_{33} & 0 & 0 & 0 \end{pmatrix} \begin{pmatrix} E_x^2 \\ E_y^2 \\ E_z^2 \\ 2E_y E_y \\ 2E_z E_x \\ 2E_x E_y \end{pmatrix},
$$

$$(2.70)$$

in the monolayer coordinate system, where $s_{14} = (\chi_{xyz}^{(2)} + \chi_{xzy}^{(2)})/2 = -(\chi_{yzx}^{(2)} + \chi_{yxz}^{(2)})/2$, $s_{15} = (\chi_{xzx}^{(2)} + \chi_{xxz}^{(2)})/2 = (\chi_{yyz}^{(2)} + \chi_{yzy}^{(2)})/2$, $s_{31} = \chi_{zxx}^{(2)} = \chi_{zyy}^{(2)}$, and $s_{33} = \chi_{zzz}^{(2)}$, s_{14}, s_{15}, s_{31}, and s_{33} are the symmetric components of second-order nonlinear susceptibilities. For the case that second-order nonlinear polarizations are generated by electric fields of two different frequencies (that are relevant to sum frequency generation and differential frequency generation), the asymmetric component of second-order nonlinear polarizabilities also contribute to the second-order nonlinear polarizations. The components s_{14}, s_{15}, s_{31}, and s_{33} of second-order susceptibilities have the form [10–13]

$$s_{14} = N_s S_2 \beta_{14}, \tag{2.71}$$

$$s_{15} = \frac{N_s}{5}(S_1 - S_3)(\beta_{33} - \beta_{31}) + \frac{N_s}{5}(3S_1 + 2S_3)\beta_{15}, \tag{2.72}$$

$$s_{31} = \frac{N_s}{5}(S_1 - S_3)(\beta_{33} - 2\beta_{15}) + \frac{N_s}{5}(4S_1 + S_3)\beta_{31}, \tag{2.73}$$

$$s_{33} = \frac{2}{5}N_s(S_1 - S_3)(\beta_{31} + 2\beta_{15}) + \frac{N_s}{5}(3S_1 + 2S_3)\beta_{33}. \tag{2.74}$$

These are derived by substituting the third term of Eq. (2.52) into Eq. (2.61), where we used Eqs. (2.59) and (2.60) to calculate the thermodynamic average. s_{15}, s_{31}, and s_{33} are achiral components that can be non-zero for both monolayers of achiral molecules and monolayers of chiral molecules, and these components are functions of orientational order parameters S_1 and S_3; second-order nonlinear polarizations arising from these achiral components are not generated from centrosymmetric bulk materials. s_{14} is proportional to the chiral component of second-order hyperpolarizability β_{14} and thus is zero for monolayers composed of achiral molecules [14]. Substituting Eq. (2.70) into Eq. (2.69) leads to second-order nonlinear polarizations in the vector form [13, 15]

$$\mathbf{P}^{(2)} = s_{31}E^2\mathbf{n} + (s_{33} - s_{31} - 2s_{15})(\mathbf{E} \cdot \mathbf{n})^2\mathbf{n} + 2s_{15}(\mathbf{E} \cdot \mathbf{n})\mathbf{E}$$
$$+ 2s_{14}(\mathbf{E} \cdot \mathbf{n})(\mathbf{E} \times \mathbf{n}). \tag{2.75}$$

Second-order nonlinear polarizations are not generated in isotropic materials because their orientational order parameters S_1, S_2, and S_3 are all zero. Monolayers that are composed of achiral molecules (and thus chiral components s_{14} are zero) show second-order nonlinear polarizations arising from achiral components, but do not show those arising from chiral components. The chiral component s_{14} is proportional to nematic order parameter S_2, see Eq. (2.71); second-order nonlinear polarizations are, in principle, generated in centrosymmetric bulk materials as long as these materials are composed of chiral molecules and show nematic order parameters, $S_2 \neq 0$.

2.4. Summary

In macroscopic electrostatics, dielectric polarizations are defined as the net displacement of electric charges and are also viewed as net electric dipoles per unit volume. In principle, macroscopic electric fields are determined when the forms of dielectric polarizations $P(r)$ are given as functions of applied electric fields. In isotropic liquids, $S_1 = 0$, $S_2 = 0$, and $S_3 = 0$, dielectric polarizations are linear to macroscopic electric fields. Electronic polarizations are due to the displacement of the center of gravity of electrons relative to the position of nuclei. Orientational polarizations are due to the competition between thermal energy and electrostatic energy arising from interactions between permanent dipoles and local electric fields. These two polarization mechanisms are particularly important in this book. In contrast to isotropic liquids, monolayers have spontaneous polarizations, linear polarizations, and second-order nonlinear polarizations, because of their non-centrosymmetric orientational structures.

Using a simple microscopic model, explicit forms of spontaneous polarizations, linear polarizations, and second-order nonlinear polarizations are shown as functions of non-zero S_1, S_2, and S_3 [16,17].

References

[1]   J. C. Maxwell, *A Treatise on Electricity and Magnetism*, Vols. 1, 2, Dover, New York, 1954.

[2]   R. P. Feynman, R. B. Lighton, and M. Sands, *The Feynman Lectures on Physics*, Vol. II. *Electromagnetism and Matter*, Basic Books, New York, 1964.

[3]   S. Sunagawa, *Theoretical Electromagnetism (Riron Denjikigaku)* (in Japanese), 3rd ed., Kinokuniya, Tokyo, 1990.

[4]   P. Debye, *Polar Molecules*, The Chemical Catalog Company, Inc., New York, 1929; Translated into Japanese by K. Nakamura and K. Sato ("YuKyokusei bunshi", Kodansha, Tokyo, 1976).

[5]   L. D. Landau and E. M. Lifshitz, *Electrodynamics of Continuous Media*, Chap. II — *Electrostatics of Dielectrics*, 2nd ed., Vol. 8 in Course of Theoretical Physics Book, Reed Education and Professional Publishing Ltd., Oxford, 1984.

[6]   S. Oka and O. Nakata, *Theory of Solid Dielectrics (Kotai Yudentai Ron)* (in Japanese), Iwanami Shoten, Tokyo, 1960.

[7]   H. Frohlich, *Theory of Dielectrics*, Clarendon Press, London, 1949. Translated into Japanese by T. Nagamiya and H. Nakai ("Yuudentai ron" Yoshioka, Tokyo, 1960).

[8]   A. J. Dekker, *Solid State Physics: Electrical Engineering Materials*, Prentice-Hall, 1958; Translted into Japanese by Y. Sakai and S. Yamanakai (*Denki Bussei Ron*, Maruzen, Tokyo, 1961).

[9]   T. Hino, *Electrical and Electronic Materials Property Engineering* (Denkizairyou Bussei Kougaku (in Japanese), Asakura, Tokyo, 1985.

[10] J. F. Nye, *Physical Properties of Crystals*, Oxford University Press, Oxford, 1985.

[11] N. Bloembergen, *Nonlinear Optics*, 4th ed., World Scientific, Singapore, 1996.

[12] Y. R. Shen, *The Principles of Non-linear Optics*, John Wiley & Sons. Inc., New York, 1984.

[13] R. W. Boyd, *Nonlinear Optics*, 3rd ed., Elsevier Inc., MA, 2008.

[14] M. Iwamoto and C. X. Wu, *The Physical Properties of Organic Monolayers*, World Scientific Publishing Ltd., Singapore, 2001.

[15] H. Fujimaki, T. Manaka, H. Ohtake, A. Tojima, and M. Iwamoto, Second-harmonic generation and Maxwell displacement current spectroscopy of chiral organic monolayers at the air–water interface, *J. Chem. Phys.*, 119, 7427–7434 (2003).

[16] M. Iwamoto, C. X. Wu, and Z. C. Ou-Yang, Second-order susceptibility tensor of a monolayer at the liquid–air interface: SHG spectroscopy by compression, *Chem. Phys. Letts.*, 325, 545–551 (2000).

[17] M. Iwamoto, T. Manaka, and Z.-C. Ou-Yang, Monolayer dielectrics and generation of Maxwell-displacement current and optical second harmonics, *IEEE Trans. Dielect. Elect. Insul.*, 11(5), 785–796 (2004).

Chapter 3

Dielectric Physics of Monolayers: Experimental Techniques to Measure Dielectric Polarizations of Monolayers

In Chapter 2, dielectric polarization of monolayers was theoretically treated. It was shown that monolayers on water surface have the characteristic dielectric polarizations that are different from bulk materials, owing to their non-centrosymmetric structures. The difference is characterized by the orientational orders that are expressed using parameters S_1, S_2, and S_3. Consequently, spontaneous polarizations, linear and nonlinear polarizations of monolayers are represented using these parameters. Therefore, experiments that can measure spontaneous, linear, and nonlinear polarizations of monolayers are helpful for studying the dielectric polarizations of monolayers. The author's group has developed an experimental system that can probe polarizations of monolayers and determine orientational order parameters. The developed experimental system consists of three parts. They are Maxwell displacement current (MDC) measuring system, Brewster angle reflectometry (BAR) system, and optical second harmonic generation (SHG) system. These three parts can measure spontaneous, spontaneous, linear, and second-order nonlinear polarizations of monolayers, respectively, and are available for determining order parameters.

3.1. Maxwell Displacement Current

Maxwell displacement current (MDC) measurement technique was first demonstrated by Iwamoto and Majima [1–4], and was employed to detect the photoisomerization of azobenzene monolayers [5] and the phase transitions of Langmuir monolayers [6, 7]. MDC is used to measure spontaneous polarizations of monolayers. A monolayer is placed between two electrodes, where one electrode is suspended in the air (electrode 1) and the other electrode is immersed in the water (electrode 2), and these two electrodes are short-circuited (see Fig. 3.1(a)). Spontaneous polarizations of monolayers induce electric charges on electrode 1 and the changes of the amount of induced charges by monolayer compression (or expansion) lead to capacitive currents (Maxwell displacement currents, MDC). That is, MDC is the transient current given by

$$I = -\frac{dQ_1}{dt}. \tag{3.1}$$

Here Q_1 is induced charge on electrode 1. The relationship between MDC currents and spontaneous polarizations is derived by using the equivalent circuit (see Fig. 3.1(b)).

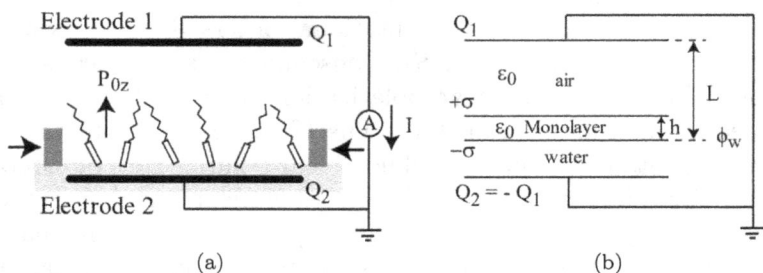

(a)   (b)

Fig. 3.1: The experimental geometry of Maxwell displacement current (MDC) measurements (a). Spontaneous polarizations $P_{0z}(= \sigma h)$ generated in monolayers (whose thickness is h) induce electric charges, Q_1 and Q_2, on the two electrodes 1 and 2. The changes of spontaneous polarizations drive capacitive currents I (Maxwell displacement currents) between these two electrodes via the external circuit. The relationship between these capacitive currents and spontaneous polarizations are analyzed by the equivalent circuit (b). We assume that the dielectric constants of air and monolayers (that are modeled as a 2D assembly of electric dipoles) are equal to that of vacuum ϵ_0. L is the distance between the water surface and electrode 1 and ϕ_w is the electrostatic potential at the water surface.

Assuming that spontaneous polarizations P_{0z} (: z-component of polarizations normal to the water surface) of monolayer is uniform and electrostatic surface potential of the water is given by ϕ_w, electric charges Q_1 is derived as

$$Q_1 = -\frac{BP_{0z}}{L} - C\phi_w, \tag{3.2}$$

where L is the distance between the electrode 1 and the water surface, B is the effective area of the electrode 1, and $C(\equiv \epsilon_0 B/L)$ is the capacitance between the water surface and electrode 1. Therefore, the form of MDCs leads to

$$I = -\frac{dQ_1}{dt} \simeq \frac{d}{dt}\left(\frac{BP_{0z}}{L}\right), \tag{3.3}$$

under the assumption that the surface potential ϕ_w is not sensitive to the changes of spontaneous polarizations during monolayer compression. The z-component of spontaneous polarizations P_{0z} generated in monolayers is determined by integrating the MDCs, under the assumption that the initial values of spontaneous polarizations are zero when the spreading area of molecules on water surface is very large, i.e., when the area per molecule is large enough. Indeed, in the region of large molecular-area A, compressing (or expanding) monolayers with a constant speed does not generate MDCs.

The orientational order parameter S_1 is determined by using the relation $P_{0z} = \mu N_s S_1$, as a function of molecular area A ($=1/N_s$). Here μ is permanent dipole, and N_s is the inverse of area per molecule A and it changes according to monolayer compressions (or expansions).

3.2.   Brewster Angle Reflectometry Microscope

BAR measures linear polarizations of monolayers. Lights are alternating electric and magnetic waves that propagate themselves through materials (and even vacuum) [8,9]. In linear optics, the optical properties of materials are characterized by refractive indices. The magnitudes of lights reflected from air–water interfaces depend on the refractive indices of air and water (see Fig. 3.2). For the cases where p-polarized lights are incident to air–water interfaces at a special

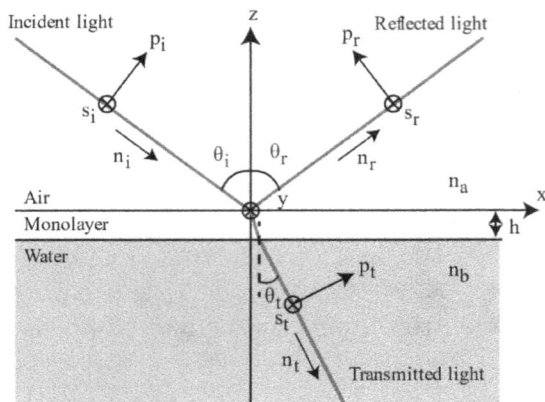

Fig. 3.2: Geometry of Brewster angle reflectometry and microscope. A monolayer at an air–water interface is irradiated by laser lights with an angle θ_i. A fraction of these lights is reflected at the interface and the other is transmitted into the water. The subscripts i, r, and t stand for incident, reflected, and transmitted lights, respectively. The unit vectors n_k ($k = i, r$, and t) are parallel to wave vectors and θ_k ($k = i, r$, and t) is the angle between the interface normal and the wave vector. The unit vectors p_k and s_k ($k = i, r$, and t) are parallel to the electric fields of p- and s-polarized lights, respectively. n_{air} and n_{wat} are the refractive indices of medium a (air) and b (water).

angle, called Brewster angle, no lights are reflected. On the other hand, when s-polarized lights are incident to monolayers on water surface at the same special Brewster angle, lights are reflected due to linear polarizations induced in monolayers.

Electric fields of incident lights have the form

$$\mathbf{E}^{\text{inc}}(\mathbf{r}) = \frac{1}{2}[\mathbf{E}_0^{\text{inc}} e^{i(\mathbf{k}_i \cdot \mathbf{r} - \omega t)} + c.c.], \qquad (3.4)$$

where $c.c.$ indicates the complex conjugate of the previous term. $\mathbf{k}_i$ is the wave vector of incident lights that have the magnitudes of $2\pi/\lambda$ (λ: wave length) and define the directions of light propagations. ω is angular frequency $2\pi f$ (f: frequency of lights). ω/k_i is the speeds of lights and depends on the refractive indices of media, in which these lights transmit ($c = 3.0 \times 10^8$ m/s in vacuum). The wavelengths of (visible) lights are about 400–750 nm and thus their frequencies are 400–750 THz. $\mathbf{r}$ is the positional vector in the air and t is time. According to Maxwell's theory of electromagnetic waves [8, 9], the directions of the electric fields $\mathbf{E}^{\text{inc}}$ of lights are perpendicular to their wave vector $\mathbf{k}_i$; the directions of the electric fields $\mathbf{E}^{\text{inc}}$ are specified by the linear combination of two independent vectors that are perpendicular to the wave vector $\mathbf{k}_i$. The directions of the electric fields of lights are called the polarizations of lights. The planes made by the wave vectors $\mathbf{k}_i$ of incident lights and the normal vectors of interfaces, to which lights are incident, are called incident plane. Two unit-vectors $\mathbf{p}_i$ and $\mathbf{s}_i$ are used to specify the directions of the electric fields of incident lights, and they are parallel and normal to the incident plane, respectively, see Fig. 3.2; $\mathbf{p}_i \times \mathbf{s}_i$ points in the direction to the wave vector $\mathbf{k}_i$. For the cases of lights, whose electric fields are directed in the direction of $\mathbf{p}_i$, these lights are called *p*-polarized and, for the cases of lights, whose electric fields are directed in the direction of $\mathbf{s}_i$, these lights are called *s*-polarized.

The wavelengths and the directions of their electric fields (polarizations) characterize electric fields. The (angular) frequencies ω of lights are independent of propagating media, whereas the wavenumber k_i is proportional to the refractive indices n of media. Maxwell's theory of electromagnetic wave predicts that the refractive indices of media are due to the linear polarizations of these media and have the form [8, 9]

$$n = \sqrt{\epsilon_r}, \tag{3.5}$$

where relative dielectric constants ϵ_r ($\equiv \epsilon/\epsilon_0$) are the dielectric constants ϵ of the media divided by the dielectric constant ϵ_0 of vacuum. It is worth to note that the relative dielectric constants of media are functions of the angular frequency ω of applied electric fields (dielectric dispersion). Because the frequencies of (visible) lights are in the order of 100 THz, ionic, orientational, and interfacial polarizations cannot follow the oscillations of applied electric fields and thus only electronic polarizations contribute to refractive indices (see Eq. (3.5)). In other words, refractive indices n of media is the values that are defined in the optical frequency region. Because of this fact, for example, the square of the refractive index of water, $n_{\text{wat}}^2 \simeq (1.33)^2 = 1.77$ is quite different from its static relative dielectric constant $\simeq 78$.

We here treat the cases that lights are incident to an air/water interface from the air (see Fig. 3.2). In these cases, in general, parts of lights are reflected to the air and the others are transmitted into the water. Reflectivity is the ratio of the magnitudes of electric fields of reflected lights to the magnitudes of the electric fields of incident lights and depends on the polarizations of incident lights. Note that reflectivity usually refers to the ratio of the intensity of reflected light to that of incident light, but, in this book, we define reflectivity as the ratio of (the magnitudes of) the electric fields of reflected light to those of incident light.

Maxwell's theory of electromagnetic waves predicts that the reflectivity of p- and s-polarized lights that are incident to air–water interfaces have the form [8]

$$R_p = \frac{n_{\text{wat}} \cos \theta_i - n_{\text{air}} \cos \theta_t}{n_{\text{air}} \cos \theta_t + n_{\text{wat}} \cos \theta_i} = \frac{\tan(\theta_i - \theta_t)}{\tan(\theta_i + \theta_t)}, \tag{3.6}$$

$$R_s = \frac{n_{\text{air}} \cos \theta_i - n_{\text{wat}} \cos \theta_t}{n_{\text{air}} \cos \theta_i + n_{\text{wat}} \cos \theta_t} = -\frac{\sin(\theta_i - \theta_t)}{\sin(\theta_i + \theta_t)}, \tag{3.7}$$

where θ_i and θ_t are incident and transmitted angles (that are angles between the interface normal and the wave vectors of incident and transmitted lights), respectively (see Fig. 3.2). n_{air} is the refractive index of medium a (air) and n_{wat} is the refractive index of water. In the derivation of Eqs. (3.6) and (3.7), relationship between the transmitted angle θ_t and the incident angle θ_i is used according to Snell's law,

$$n_{\text{air}} \sin \theta_i = n_{\text{wat}} \sin \theta_t. \tag{3.8}$$

Because the refractive index of water is larger than that of air, $n_{\text{wat}} > n_{\text{air}}$, transmitted angles are smaller than incident angles, $\theta_t < \theta_i$ for the case that light is incident from the air (see Eq. (3.8)). This implies that reflected light arising from s-polarized incident light does not vanish for any incident angles θ_i (see Eq. (3.7)). In contrast, reflected light arising from p-polarized incident light vanishes at an incident angle $\theta_i = \theta_B$, where $\theta_B + \theta_t = \pi/2$ (see Eq. (3.6)). This angle θ_B is a special angle called Brewster angle and has the relationship

$$\tan \theta_B = \frac{n_{\text{wat}}}{n_{\text{air}}}, \tag{3.9}$$

(see Eq. (3.8) with $\theta_i = \theta_B$ and $\theta_B + \theta_t = \pi/2$). The direction of electric fields of s-polarized light is parallel to the interface for any incident angles. By contrast, the direction of electric fields of

p-polarized light changes according to the change of incident angle and thus, at Brewster angle, the boundary condition of electromagnetic fields holds even without reflected light; p-polarized incident lights have Brewster angle $\theta_i = \theta_B$, at which reflected lights vanish but s-polarized lights do not. The Brewster angle of air–water interfaces, $n_{\mathrm{air}} \simeq 1$ and $n_{\mathrm{wat}} \simeq 1.33$, is about 53.1°. The fact that reflected lights of p-polarized incident light vanish at the Brewster angle θ_B implies that the contributions of linear polarizations generated in air and water to reflected lights are eliminated [8, 9]. Consequently, reflected light in this condition is sensitive to linear polarizations generated in monolayers at the air–water interface [10, 11].

We here treat monolayers that show C_∞-symmetry, with the thickness h, and their dielectric properties are specified using the dielectric constant tensor of the form

$$\epsilon = \begin{pmatrix} \epsilon_\| & 0 & 0 \\ 0 & \epsilon_\| & 0 \\ 0 & 0 & \epsilon_\perp \end{pmatrix}. \tag{3.10}$$

Here the components of this dielectric constant tensor are represented using the monolayer coordinate. Noteworthy that the dielectric constant tensor, Eq. (3.10), leads to 3D linear polarizations that are dipoles per unit volume, whereas monolayers are 2D polarizations that are dipoles per unit area. Dielectric constant tensor, Eq. (3.10), has approximate relationships with linear susceptibility $\chi^{(1)}$, $\chi_\perp^{(1)} = (\epsilon_\perp - \epsilon_0)h$ and $\chi_\|^{(1)} = (\epsilon_\| - \epsilon_0)h$; our treatment of thin 3D films is thus only an approximate treatment. These relationships and Eqs. (2.65) and (2.66) imply that the components of dielectric constant tensor $\epsilon_\perp$ and $\epsilon_\|$ have the form

$$\epsilon_\perp = \frac{2}{3}\Delta\epsilon S_2 + \bar{\epsilon}, \tag{3.11}$$

$$\epsilon_\| = -\frac{1}{3}\Delta\epsilon S_2 + \bar{\epsilon}, \tag{3.12}$$

$\epsilon_\perp$ and $\epsilon_\|$ are functions of the orientational order parameter S_2. The wave length λ of light is in the order of 450–700 nm and is much longer than the thickness of monolayers h ($\sim 1\,\mathrm{nm}$). By the first-order term with respect to h/λ, the reflectivity r_p has the form [12, 13]

$$r_p = R_p + \Delta r_p, \qquad (3.13)$$

with

$$\Delta r_p = ikh \frac{2n_{\text{air}} \cos \theta_i \cos^2 \theta_t}{(n_{\text{wat}} \cos \theta_i + n_{\text{air}} \cos \theta_t)^2} \left[n_{\text{air}}^2 + n_{\text{wat}}^2 - \epsilon_\parallel - \frac{n_{\text{air}}^2 n_{\text{wat}}^2}{\epsilon_\perp} \right].$$

$$(3.14)$$

k is the wave number of (incident) light in vacuum. The values of the orientational order parameter S_2 are determined by using Eq. (3.14). Equation (3.13) is the general equation that can be used for any incident angles θ_i, but the use is limited for the case of monolayers that show C_∞-symmetry. The second term of Eq. (3.13), Δr_p, results from linear polarizations generated from monolayers and this term is highlighted at the Brewster's condition, $R_p = 0$.

BAR is an optical technique that measures lights reflected from monolayers and can be used as a microscopic visualization technique, e.g., Brewster-angle microscope (BAM) [10,11]. BAM has been widely used to observe domains in monolayers [10–13]. In contrast to fluorescence microscopes that are another widely used as visualization techniques, BAM does not use fluorescent probes that potentially affect the structures of monolayers. Moreover, reflected lights that are corrected by BAR and BAM include the information of linear polarizations and thus analyzing Brewster angle micrographs provide the information of orientations of constituent molecules. BAM visualizes monolayers as profiles of refractive indices at air–water interfaces (see Eq. (3.14)). The shapes of domains in monolayer can be visualized by using BAM, because, in many cases, domains of different phases have different values of refractive indices.

3.3.   Optical Second Harmonics Measurements

When lights are incident on so-called nonlinear optical materials (monolayers are one of them), these materials generate second harmonic lights that have double frequency of the incident lights. These phenomena are called optical second harmonic generations (SHG) [14–16]. Optical SHG can be used to measure second order nonlinear polarizations of monolayers [12–18]. In general, nonlinear

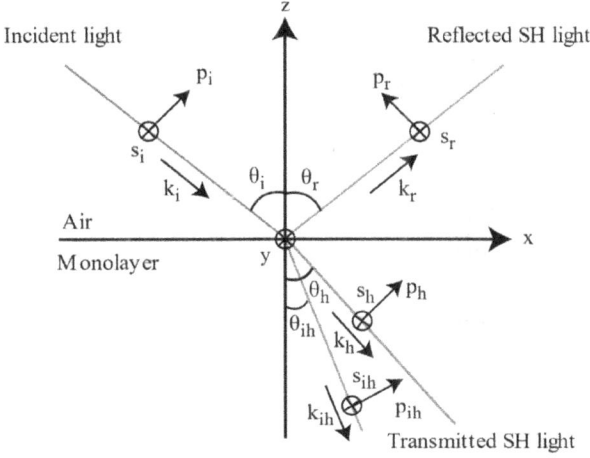

Fig. 3.3: Geometry of optical SHG. Fundamental lights are incident to mono-layers at an incident angle θ_{inc} and the electric fields of these fundamental lights generate second harmonic lights. The subscripts ref and tra stands for reflected and transmitted second harmonic lights, respectively. $\mathbf{p}_k$ and $\mathbf{s}_k$ are the unit vectors to the electric fields of p- and s-polarized incident fundamental lights, $k = $ inc, reflected second harmonic lights, $k = $ ref, and transmitted second harmonic lights, $k = $ tra. $\mathbf{k}_k$ are the wave vectors of incident fundamental lights, $k = $ inc, reflected second harmonic lights, $k = $ ref, and transmitted second harmonic lights, $k = $ tra and θ_k are the angles between the wave vector $\mathbf{k}_k$ and the interface normal with $k = $ inc, ref, and tra.

polarizations generated from monolayers are relatively small com-pared to spontaneous and linear polarizations, and thus high-power laser beams are employed in experiments. We here analyze the cases that lights with an angular frequency of ω are incident on monolayers (see Fig. 3.3).

The electric fields of incident lights in monolayers have the form

$$\mathbf{E}^{inc}(\mathbf{r}) = \frac{1}{2}\mathbf{E}_0^{inc}(\mathbf{r})e^{-i\omega t} + c.c., \tag{3.15}$$

where *c.c.* is the complex conjugate of the first term. These electric fields generate second-order nonlinear polarizations in monolayers, which have the form

$$\mathbf{P}^{(2)}(\mathbf{r}) = \frac{1}{4}[\chi^{(2)}: \mathbf{E}_0(\mathbf{r})\mathbf{E}_0(\mathbf{r})e^{-2i\omega t} + c.c.]$$

$$+ \frac{1}{4}[\chi^{(2)}: \mathbf{E}_0^*(\mathbf{r})\mathbf{E}_0(\mathbf{r}) + c.c.]. \tag{3.16}$$

Second-order nonlinear polarizations have terms that oscillate with an angular frequency 2ω $(= \omega + \omega)$, see the first term of Eq. (3.16). Noteworthy that the second term of Eq. (3.16) represents the nonlinear polarizations generated at zero angular frequency 0 $(= \omega - \omega)$, and this *d.c.* nonlinear polarization is also generated by the electric fields of incident lights with angular frequency ω. However, we do not treat the *d.c.* nonlinear polarization. There are standard text books on the electromagnetic theory of optical second harmonic generation (SHG) [14–16] and we do not outline the theory here. On the other hand, we here phenomenologically utilize the concept of the physics of SHG that second-order nonlinear polarizations oscillating with an angular frequency of 2ω generate second harmonic lights that have an angular frequency of 2ω.

The intensities of second harmonic lights have the form

$$I_{sh} \propto |\mathbf{e}_{sh} \cdot \chi^{(2)} : \mathbf{e}_{inc}\mathbf{e}_{inc}|. \qquad (3.17)$$

The unit vector $\mathbf{e}_{inc}$ is parallel to the electric fields of fundamental (incident) lights, $\mathbf{E}_{inc}(\mathbf{r}) = |\mathbf{E}_{inc}(\mathbf{r})|\mathbf{e}_{inc}$. In many experiments, we selectively detect the components of (the electric fields) of second harmonic lights to a certain direction by using polarizers and $\mathbf{e}_{sh}$ is the unit vector to this direction. It is important to note that $\mathbf{e}_{sh}$ is different in directions for reflected and transmitted second harmonic lights because the propagating directions of these second harmonic lights are different.

Thus far, our discussion is limited to the case of dielectric properties of monolayers that have C_∞-symmetry (see Section 2.3 in Chapter 2). Because second-order susceptibilities $\chi^{(2)}$ reflect the symmetry of monolayers, one can know the symmetry of the system by measuring second harmonic lights. Moreover, second-order nonlinear polarizations depend on the orientational order parameters S_1 and S_3 and thus, for the case of monolayers that have C_∞-symmetry, the values of S_1 and S_3 can be determined by measuring second harmonic lights.

3.3.1. *Symmetry of monolayers and SHG*

The symmetry of monolayers can be determined by measuring the intensity of second harmonic lights for a set of four fundamental polarizations, p–p, p–s, s–p, and s–s. Here p–s stands for the s-components of second harmonic lights that are generated by p-polarized fundamental lights, etc.

For monolayers that have C_∞-symmetry, second-order nonlinear polarizations have the form of Eq. (2.71) (see Chapter 2). The intensities of transmitted second harmonic lights for the four fundamental sets of polarizations have the forms

$$I_{p-p} \propto |(s_{31} \cos^2 \theta_{\rm inc} + s_{33} \sin^2 \theta_{\rm inc}) \sin \theta_{\rm tra}$$

$$+ 2s_{15} \sin \theta_{\rm inc} \cos \theta_{\rm inc} \cos \theta_{\rm tra}|^2, \tag{3.18}$$

$$I_{p-s} \propto |2s_{14} \sin \theta_{\rm inc} \cos \theta_{\rm inc}|^2, \tag{3.19}$$

$$I_{s-p} \propto |s_{31} \sin \theta_{\rm tra}|^2, \tag{3.20}$$

$$I_{s-s} = 0, \tag{3.21}$$

(see Eq. (3.17)), where $\theta_{\rm inc}$ and $\theta_{\rm tra}$ are incident and transmitted angles, respectively (see Fig. 3.3). The intensities of reflected second harmonic lights are given by replacing $\theta_{\rm tra}$ with $\pi - \theta_{\rm ref}$, where $\theta_{\rm ref}$ is reflected angle. The intensity I_{p-s} at the p–s polarization is proportional to the chiral component s_{14} of second order susceptibility. This implies that the chirality of monolayers is detected by measuring second harmonic lights in the p–s polarization. Second harmonic lights in the s–s polarizations are not generated from monolayers of C_∞-symmetry. In many cases, monolayers show C_∞-symmetry when constituent molecules do not show in-plane orientational ordering.

For the cases of monolayers, where their constituent molecules show in-plane orientational ordering toward the x-direction, see Fig. 3.3, these monolayers generate second-order nonlinear polarizations that have the form

$$\begin{pmatrix} P_x^{(2)} \\ P_y^{(2)} \\ P_z^{(2)} \end{pmatrix} = \begin{pmatrix} s_{11} & s_{12} & s_{13} & 0 & s_{15} & 0 \\ 0 & 0 & 0 & s_{24} & 0 & s_{26} \\ s_{31} & s_{31} & s_{33} & 0 & s_{35} & 0 \end{pmatrix} \begin{pmatrix} E_x^2 \\ E_y^2 \\ E_z^2 \\ 2E_y E_y \\ 2E_z E_x \\ 2E_x E_y \end{pmatrix}. \tag{3.22}$$

The form of this second-order susceptibility is derived by using the fact that these monolayers have the mirror symmetry with respect to z–xx plane, C_s-symmetry, and this is the cases of monolayers that are composed of achiral molecules. The intensities of second harmonic lights for the four fundamental polarization sets have the forms

$$I_{p-p} \propto j(s_{31} \cos^2 \theta_{\mathrm{inc}} + s_{33} \sin^2 \theta_{\mathrm{inc}}) \sin \theta_{\mathrm{tra}}$$

$$+ 2s_{24} \sin \theta_{\mathrm{inc}} \cos \theta_{\mathrm{inc}} \cos \theta_{\mathrm{tra}}|^2. \tag{3.23}$$

$$I_{p-s} \propto |s_{12} \cos^2 \theta_{\mathrm{inc}} + s_{13} \sin^2 \theta_{\mathrm{inc}}|^2, \tag{3.24}$$

$$I_{s-p} \propto |s_{31} \sin \theta_{\mathrm{ref}}|^2 \tag{3.25}$$

$$I_{s-s} \propto |s_{11}|^2. \tag{3.26}$$

Second harmonic lights in the s–s polarization are thus generated from the monolayers of C_s-symmetry; whether a monolayer has C_∞-symmetry or not can be determined by measuring second harmonic lights in the s–s polarization. The results of the analysis of the four fundamental polarization sets are summarized in Table 3.1.

3.3.2.   *Determination of orientational order parameters S_1 and S_3*

For the cases of rod-shaped molecules that have long conjugated units along their long-axes (and thus their π-electrons can displace in their long-axis much longer distance than in their short-axes), the ww-component β_{33} of second-order hyperpolarizability dominates the other components (see Fig. 2.5 in Chapter 2). Monolayers composed of these rod-shaped molecules show second order susceptibilities, whose components have simplified forms (see Eqs. (2.73)–(2.75) for

Table 3.1:   Second harmonic lights for the four fundamental polarization sets, p–p, p–s, s–p, and s–s, where p–s stands for s-component of second harmonic lights generated by p-polarized fundamental lights, etc.

	p–p	p–s	s–p	s–s
C_∞ (chiral)	Y	Y	Y	N
$C_{\infty v}$ (achiral)	Y	N	Y	N
C_s	Y	Y	Y	Y

Note: Y indicates the cases that second harmonic lights are generated and N indicates the cases that these lights are not generated.

$\beta_{33} \gg \beta_{15}, \beta_{31}, \beta_{14}$ in Chapter 2). The arbitrary directions of the polarization of fundamental lights and second harmonic lights are represented by the linear combination of the unit vector of p- and s-polarization in the forms

$$\mathbf{e}_k = \cos \gamma_k \mathbf{p}_k + \sin \gamma_k \mathbf{s}_k, \tag{3.27}$$

with $k = i$, r, and t, where γ_k, $k = $ inc, ref, and tra, are polarization angles (see also Fig. 3.3). The subscript tra stands for transmitted second harmonic lights. The components of second harmonic lights in $\mathbf{e}_k$ have the form [17]

$$I_{\text{ref}} \propto |A_{\text{refl}} S_1 + B_{\text{refl}} S_3|^2 I_{\text{ref}}, \tag{3.28}$$

$$I_{\text{tra}} \propto |A_{\text{tra}} S_1 + B_{\text{tra}} S_3|^2 I_{\text{tra}}, \tag{3.29}$$

in the reflected and transmitted geometries, respectively. The explicit forms of A_{ref}, B_{ref}, A_{tra}, and B_{tra} are shown in Refs. [17] and [18] (a simple approach that does not take into account the refractive indices of air and water is used in Ref. [17–21] and more elaborate approach that takes into account these refractive indices is used in Ref. [18]). A_{ref}, B_{ref}, A_{tra}, and B_{tra} are functions of polarization angles γ_{inc}, γ_{ref}, and γ_{tra}, etc., and there exist polarization angles γ_{inc}, γ_{ref}, and γ_{tra}, where either of A_{ref}, B_{ref}, A_{tra}, or B_{tra} are zero. The orientational order parameter S_1 is experimentally determined by measuring second harmonic lights for a set of polarization angles that satisfy $B_{\text{ref}} = 0$ or $B_{\text{tra}} = 0$, whereas the orientational order parameter S_3 is determined by measuring second harmonic lights in a set of polarization angles that satisfy $A_{\text{ref}} = 0$ or $A_{\text{tra}} = 0$.

3.4. Measurements of Dielectric Polarizations of Monolayers Composed of Rod-Shaped Achiral Molecules

Alkyl-cyanobiphenyl molecules are amphiphilic molecules that show nematic liquid crystal phases (see Fig. 3.4). MDC-SHG measurements [17–21] were conducted during compression of 8CB monolayers at a compression rate 4.2 Å^2/min (that is the decrease of area per molecule in a unit time $-\Delta A/\Delta t$), (Fig. 3.5, [15, 16]). The details of

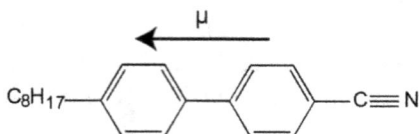

Fig. 3.4: The molecular structure of p-octyl-p'-cyanobiphenyl (8CB). 8CB molecules isotropic-nematic transitions at 41°C, nematic-smectic A transitions at 34°C, and smectic A — crystal transitions at 22°C [19]. 8CB has permanent dipole moment of ∼4.7 D toward its long-axis and second order hyperpolarizability, whose ww-component βww is ∼6.2 [C · m^3/V^2] [24].

experimental procedures are in Ref. [17,18]. These graphs are divided into three regions based on the result of surface pressure, see the bottom graph in Fig. 3.5. Second harmonic lights in the p–p and s–p polarizations are detected from the 8CB monolayer, see Fig. 3.5, the first graph from the top, meanwhile the p–s and s–s polarizations are not, the second graph from the top; results indicate that the monolayer has $C_{\infty}v$-symmetry for all regions of molecular area, and implies that orientational order parameters S_1 and S_3 can be determined by using optical second harmonic generation (see Section 3.3.2).

The polarizations of fundamental lights and second harmonic lights are set so that the intensities of reflected and transmitted second harmonic lights are proportional to the square of the orientational order parameters S_1 and S_3, respectively; the intensities of second harmonic lights in this condition are shown in Fig. 3.6, the first and second graphs. The z-component of dipoles is proportional to the orientational order parameter S_1 (see Section 2.3). The experiments for Figs. 3.5 and 3.6 were performed with the same barrier speed.

The surface pressure of the 8CB monolayer was negligibly small in region 1. This implies that the monolayer was gas and/or liquid expanded phase. MDC was negligibly zero until the monolayer was compressed to a threshold molecular area, ∼68 Å^2/molecule, see region 1A in Fig. 3.6, the second graph from the bottom. This implies that the dipoles of constituent molecules were parallel to the monolayer plane, $S_1 \simeq 0$, see region 1A in Fig. 3.5, the third graph from the bottom. This is also supported by the measurements of second harmonic lights that showed that orientational order parameters S_1 and S_3 are zero in region 1A, see Fig. 3.6, the first and second

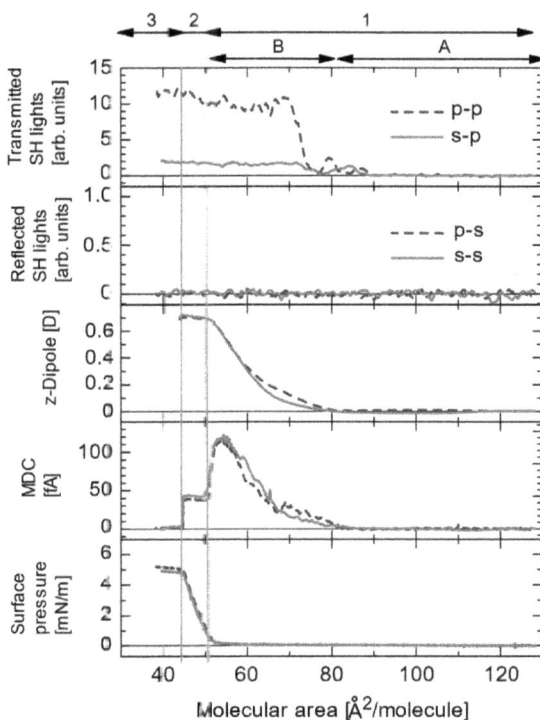

Fig. 3.5: A typical result of MDC-SHG experiments on a 8CB monolayer during a compression with a compression rate of 4.2 $\text{Å}^2/\text{min}$. The surface pressure of 8CB monolayers is shown in the first graph from the bottom. Maxwell displacement currents (MDC) and the z-component of dipole moments that were calculated from MDC results are shown in the second and third graphs, respectively. The intensities of second harmonic lights were measured for p–p, p–s, s–p, and s–s polarization sets, where p–s polarization stands for s-polarized second harmonic lights generated by p-polarized fundamental lights, the fourth and fifth graphs.

Source: Reproduced from Ref. [15], *Rev. Sci. Instrum.*, 74, 2828–2835, 2003.

graphs from the top. MDC increased gradually for the molecular area that was smaller than this threshold molecular area until MDC reached a peak, see Fig. 3.6, the second graph from the bottom. The z-component of dipoles that is proportional to the orientational order parameter S_1 increased with decreasing the molecular area. The orientational order parameters S_1 and S_3 that were determined by using

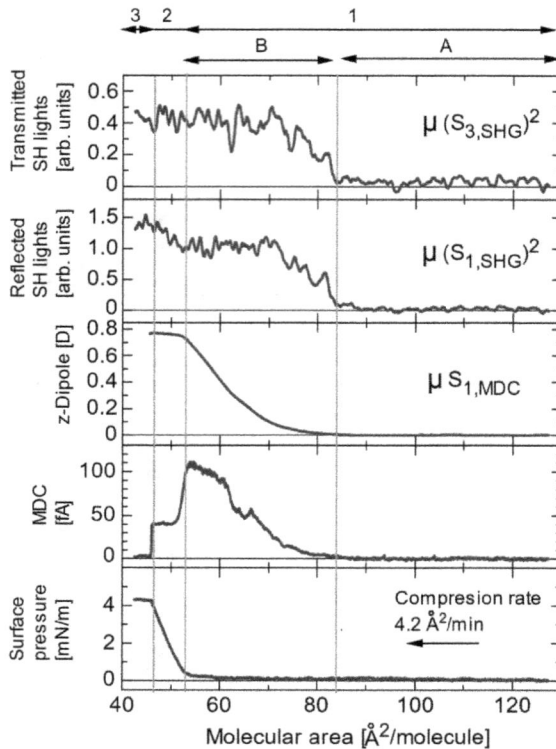

Fig. 3.6:   A typical result of MDC-SHG experiments on 8CB monolayers during a compression with a compression rate of 4.2 Å^2/min. The surface pressure, MDC, and the z component of dipole moments were shown in the first, second, and third graph from the bottom. The polarizations of fundamental and second harmonic lights were set so that the intensities of second harmonic lights were proportional to the square of $S1$ and $S3$, respectively, see the first and second graphs from the top.

Source: Reproduced from Ref. [18], *Rev. Sci. Instrum.*, 74, 2828–2835, 2003.

optical second harmonic generation showed a similar tendency, see Fig. 3.6, the first and second graphs from the top.

The surface pressure gradually increased in region 2, see Fig. 3.5, the first graph from the bottom. This implies that the monolayer was condensed phase in this region. MDC increased only slightly in region 2; the dipoles of the constituent molecules did not change their orientations much. The orientational order parameter S_1 that was determined by using second harmonic lights increased slightly and the orientational order parameter S_3 was almost constant in this region. The surface pressure is saturated in region 3 (see Fig. 3.5). MDC abruptly changed to zero; the average orientation of constituent molecules did not change while the monolayer was compressed. This implies that constituent molecules were pushed out from the monolayer and stabilized a multilayer when this monolayer was compressed even at such a large surface pressure. The observation by using Brewster angle microscopes gives more information on the structures of the molecular film in this region, see the discussion below.

MDC-BAR [12,22] experiments were performed during a compression of 8CB monolayers by compression rate of 5.6 $\text{Å}^2/\text{min}$ as shown in Fig. 3.7 [12]. The graph was divided into three regions in a similar manner to Figs. 3.5 and 3.6. The surface pressure and MDC were measured to correspond to the results of MDC-SHG measurements [17,18] and are shown in Fig. 3.7 (the first and second graphs from the bottom). The z-component of dipoles is shown in the third graph from the bottom to analyze the result. The values of MDC in Fig. 3.7 are larger than the values of MDC in Fig. 3.5 by the factor of 1.4; this is because of the fact that we used larger compression rate (by the factor of 1.3) for the MDC-BAR experiments. The intensities of reflected lights are almost constant throughout region 1, see Fig. 3.7, the first graph from the top. Because the results of MDC-SHG measurements suggest that the long-axes of constituent molecules are parallel to the monolayer plane in region 1A, we assumed that the orientational order parameter S_2 was zero in this region.

The intensities of reflected light increased gradually in region 2. The orientational order parameter S_2 was estimated. This estimates that the orientational order parameter S_2 jumps from zero to 0.7 at 55 $\text{Å}^2/\text{mol}$, where the monolayer showed a phase transition (from liquid expanded phase and condensed phase), and increased slightly with decreasing area per molecule.

The 8CB monolayer was uniform in both regions 1 and 2 (see Fig. 3.7 (right)). The 8CB monolayer stabilized small circular domains that coexist with darker background domains in region 3 (see Fig. 3.7). These are probably the domains of multilayer phase

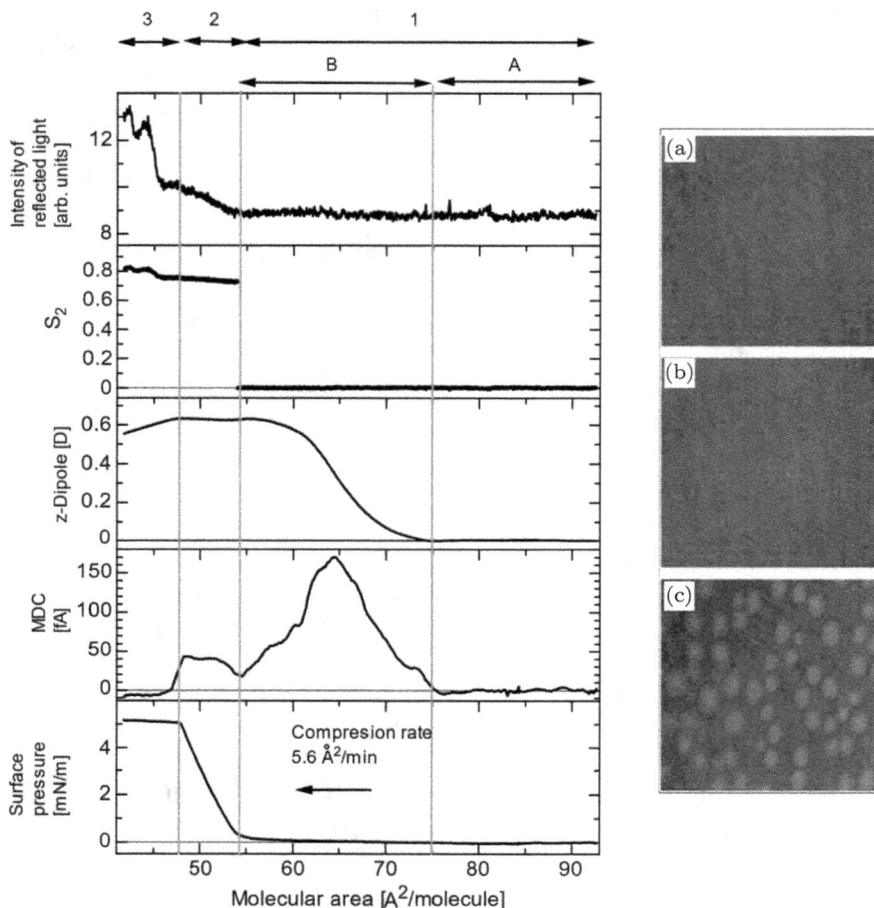

Fig. 3.7: A typical result of MDC-BAR experiments on 8CB monolayers in a compression with compression rate 5.6 Å^2/min (left). The first, second, and third graphs from the bottom are surface pressure, Maxwell displacement currents (MDC), and the z-component of dipoles. The first graph from the top is the intensities of reflected lights at Brewster's angle. Reflectivity was used to determine the orientational order parameter S_2 and the second graph from the top is proportional to S_2. Brewster angle micrographs at the region 1, 2, and 3 (right). *Source*: Reproduced from Ref. [12], *Rev. Sci. Instrum.*, 76, 083902 (7 p.), 2005.

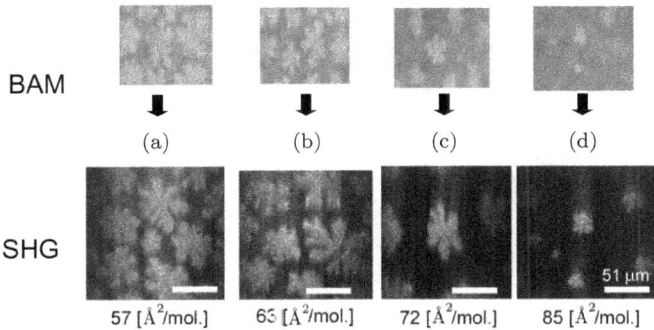

Fig. 3.8: BAM and SHG images. Visualized polarization structure of lipid Langmuir monolayer by surface second-harmonic generation.
Source: Reprinted from Ref. [22], *Thin Solid Films*, 554, 8–12, 2014.

and their size is a few micrometers. MDC-BAR-SHG experiments have demonstrated that monolayers indeed show characteristic dielectric polarizations, spontaneous polarizations, linear polarizations, and second order nonlinear polarizations.

3.5. BAM and SHG Images

Visualized images of monolayers are very useful to study physics of domain shapes. There are two directions to observe domains. One is to focus on the contours of domain shapes, where BAM images are very useful. The other is to focus on the inner texture of domains, where SHG images are more helpful. BAM can provide the images of linear polarization of domains, whereas SHG can provide the images of nonlinear polarizations of domains. Therefore we may anticipate that contours and inner textures of domains will be visualized by BAM and SHG images, accordingly. Figure 3.8 shows the BAM and SHG images of domains in NBD-PE monolayers [22], where we can see similar contours of domain shapes at various surface-pressures. Figure 3.9 shows the observed textures in domains by BAM and SHG images, where L-DPPC:NBD-PE mixed monolayer is visualized as BAM and SHG images. Texture of helical-structure is clearly

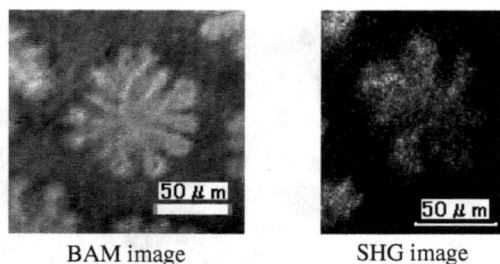

BAM image   SHG image

Fig. 3.9:   Images of 3:7 L-DPPC: NBD-PE mixed monolayer by BAM and SHG. Texture of helical-structure is observed in domains by SHG (right-hand), but it is not clear by BAM (left hand).
Source: Reprinted from Ref. [22], *Thin Solid Films*, 554, 8–12, 2014.

observed in domains by SHG (right-hand). In this way, by focusing on linear and nonlinear polarizations of monolayers, not only domain shapes but also the inner texture of domains can be visualized. These optical techniques are useful for the verification of theories and analyses.

3.6.   Summary

In macroscopic electrostatic, dielectric polarizations are defined as the net displacement of electric charges and are also viewed as net electric dipoles per unit volume. In principle, macroscopic electric fields are determined (and thus problems of electrostatics are solved) when the forms of dielectric polarizations $\mathbf{P}(\mathbf{r})$ are given as functions of applied electric fields. In isotropic liquids, $S_1 = 0$, $S_2 = 0$, and $S_3 = 0$, dielectric polarizations are linear to macroscopic electric fields and there are four typical mechanisms; electronic polarizations, ionic polarizations, orientational polarizations, and interfacial polarizations. Electronic polarizations are due to the displacement of the center of gravity of electrons relative to the position of nuclei and orientational polarizations are due to the competition between thermal energy and electrostatic energy arising from interactions between permanent dipoles and local electric fields. In contrast to isotropic liquids, $S_1 = 0$, $S_2 = 0$, and $S_3 = 0$, monolayers can generate spontaneous polarizations, linear polarizations, and second-order nonlinear polarizations, because of their non-centrosymmetric orientational structures, $S_1 \neq 0$, $S_2 \neq 0$, and $S_3 \neq 0$. We thus used a simple microscopic model to derive explicit forms of spontaneous polarizations, linear polarizations, and second-order nonlinear polarizations as functions of S_1, S_2 and S_3. Maxwell displacement current (MDC), Brewster angle reflectometry (BAR), and optical second harmonic generation (SHG) are experimental techniques to measure spontaneous, linear, and second-order nonlinear polarizations, respectively. Using the developed MDC-BAR-SHG system [12, 17, 18], generation of spontaneous, linear, and second-order nonlinear polarizations generated in monolayers is demonstrated. The fact that monolayers generate spontaneous polarizations implies that monolayers spontaneously store electrostatic energy and this indeed plays an important role in the shapes of domains in monolayers (see Chapters 5, 6, and 7).

References

[1] Y. Majima and M. Iwamoto, A new displacement current measuring system coupled with the Langmuir film technique, *Rev. Sci. Instrum.*, 62, 2228–2233 (1991).

[2] M. Iwamoto, Y. Majima, Investigations of the dynamic behavior of fatty-acid monolayers at the air-water-interface using a displacement current-measuring technique coupled with the Langmuir-film technique, *J. Chem. Phys.*, 94(7), 5135–5142, (1991).

[3] M. Iwamoto, Y. Majima, Determination of the dipole-moment of a monolayer at the air water interface using a current-measuring technique, *Jpn. J. Appl. Phys.*, 27(5), 721–725, (1988).

[4] Y. Majima, M. Iwamoto, Studies on the dynamic behavior of fatty-acid monolayers at the air-water-interface by a current-measuring technique, *Jpn. J. Appl. Phys.*, 29(3), 564–568, (1990).

[5] M. Iwamoto, Y. Majima, H. Naruse, T. Noguchi, and H. Fuwa, Generation of Maxwell displacement current across an azobenzene monolayer by photoisomerization, *Nature*, 353, 645–647 (1991).

[6] M. Iwamoto, T. Kubota, and M. R. Muhamad, Detection of phase transitions in liquid crystals on a water surface by a Maxwell displacement current measuring technique, *J. Chem. Phys.*, 102, 9368–9374 (1995).

[7] M. Iwamoto, T. Kubota, and Z. C. Ou-Yang, Maxwell displacement current across phospholipid monolayers due to phase transition, *J. Chem. Phys.*, 104, 736–741 (1996).

[8]   R. E. Collin, *Field Theory of Guided Waves*, 2nd ed., Wiley-IEEE Press, New York, 1991.

[9]   J. A. Stratton, *Electromagnetic Theory*, McGraw-Hill Book Company, New York and London, 1941.

[10] S. Henon and J. Meunier, Microscope at the Brewster angle: Direct observation of first-order phase transitions in monolayers, *Rev. Sci. Instrum.*, 62, 936–939 (1991).

[11] D. Hönig and D. Möbius, Direct visualization of monolayers at the air-water interface by Brewster angle microscopy, *J. Phys. Chem.*, 95, 4590–4592 (1991).

[12] R. Wagner, T. Yamamoto, T. Manaka, and M. Iwamoto, Determination of the complete dielectric polarization of Langmuir monolayers, *Rev. Sci. Instrum.*, 76, 083902 (2005).

[13] Y. Tabe and H. Yokoyama, Fresnel formula for optical anisotropic Langmuir monolayers: An application to Brewster angle microscope, *Langmuir*, 11, 609–704 (1995).

[14] N. Bloembergen, *Nonlinear Optics*, 4th ed., World Scientific, Singapore, 1996.

[15] Y. R. Shen, *The Principles of Non-Linear Optics*, John Wiley & Sons, Inc., New York, 1984.

[16] R. W. Boyd, *Nonlinear Optics*, 3rd ed., Elsevier Inc., MA, 2008.

[17] A. Tojima, T. Manaka, and M. Iwamoto, Orientational order study of monolayers at the air-water interface by Maxwell-displacement current and optical second harmonic generation, *J. Chem. Phys.*, 115, 9010–9017 (2001).

[18] A. Tojima, T. Manaka, and M. Iwamoto, Instrument equipped with Maxwell displacement current and optical second harmonic generation measurement system, *Rev. Sci. Instrum.*, 74, 2828–2835 (2003).

[19] H. Fujimaki, T. Manaka, H. Ohtake, A. Tojima, and M. Iwamoto, Second-harmonic generation and Maxwell displacement current spectroscopy of chiral organic monolayers at the air-water interface, *J. Chem. Phys.*, 119, 7427–7434 (2003).

[20] M. Iwamoto, C. X. Wu, and Z. C. Ou-Yang, Second-order susceptibility tensor of a monolayer at the liquid-air interface: SHG spectroscopy by compression, *Chem. Phys. Letts.*, 325, 545–551 (2000).

[21] M. Iwamoto, T. Manaka, and Z.-C. Ou-Yang, Monolayer dielectrics and generation of Maxwell-displacement current and optical second harmonics, *IEEE Trans. Dielect. Elect. Insul.*, 11(5), 785–796 (2004).

[22] Y. Matsuoka, D. Taguchi, T. Manaka, and M. Iwamoto, Visualizing polarization structure of lipid Langmuir monolayer by surface second harmonic generation technique, *Thin Solid Films*, 554, 8–12 (2014).

[23] A. Tojima, Characterization of dielectric polarizations generated from monolayers by using Maxwell displacement current and optical second harmonic generation, Ph D. thesis, Tokyo Institute of Technology (in Japanese).

[24] Liquid Crystal Handbook Editorial Committee, *Liquid Crystal Handbook*, Maruzen, Tokyo, 2000 (in Japanese).

Chapter 4

Liquid Crystalline Properties of Monolayers on Water Surface

Monolayers on water surface show behaviors that are analogous to hexatic liquid crystals (see Chapter 1), and the continuum theory of nematic liquid crystals is one of the important methods for analyzing domains in monolayers. However, monolayers are basically different from bulk LCs due to their presence on the water surface. In this chapter, on paying attention to dielectric polarizations, the liquid crystalline property of monolayers on the water surface is broadly treated based on experimental results reported. In Section 4.1, it is pointed out that monolayers comprised of amphiphilic molecules possess both the lattice-structure property related to the positional ordering of polar heads on the water surface and the liquid crystalline property related to the orientational ordering of chains of long tails of amphiphiles pointing toward air. In Section 4.2, the liquid-crystalline nature of monolayers is reviewed, and dielectric polarizations of monolayers are treated using *directors* that are defined along ordered long tails of amphiphiles. In Section 4.3, Anisotropic 2D compressibility is considered an extension of the hydrodynamic theory of bulk nematic liquid crystals by using directors, and flow-induced reorientations of monolayers are treated with consideration of the flow of a 2D hexatic structure of molecular heads on the water surface. In Section 4.4, the flexoelectric effect of monolayers is treated. Section 4.5 is the summary of this chapter.

4.1. Condensed Phases of Monolayers on Water Surface

Monolayers have been extensively investigated for more than one century since Langmuir's pioneering works. Through traditional surface pressure-area isotherm measurements, abundant phenomena of polymorphism and phase transitions in monolayers of amphiphilic molecules have been revealed, e.g., amphiphile monolayers at the air-water interface show condensed phases at relatively high surface pressures, see Fig. 1.3 (Chapter 1). Experiments on fatty acid monolayers by application of grazing incidence X-ray diffraction (GIXD) method into the observation of molecular packing structure and advanced optical observation microscopy have elucidated the presence of many phases that are characterized by orientational and translational orders. Results are summarized in many articles, e.g., in the review article by Kaganer *et al.* [1].

Interestingly, condensed phases of amphiphile monolayers are divided into *untilted* (condensed) phases, *CS*, *S*, and *LS*, at higher pressures and *tilted* (condensed) phases, L_2, L'_2, L''_2, Ov, at lower surface pressures (the structures of these phases are summarized in Fig. 4.1). In both *tilted* and *untilted* phases, the heads of constituent molecules are aligned in hexagonal lattices and these translational orders are only short-ranged (in the order of 10 molecules). In *tilted* phases, constituent molecules are tilted in either nearest neighbor (*NN*) or next nearest neighbor (*NNN*) directions and these *in-plane* orientational orders are long- ranged. This fact implies that even though the translational order of molecular heads is only short-ranged, the orientations of hexatic lattices that are specified by their *NN*-directions show long-range ordering (hexagonal order). Indeed, the hexagonal lattice of molecular heads in many of the condensed phases (except for *LS* phase) are deformed in either *NN* or *NNN* directions as a result of the tilting of molecular chains and/or the ordering of the backbone plane of the hexatic lattice of molecules [1]. The structures of condensed phases are analogous to hexatic phases of liquid crystals-hexatic B, I, and F. These facts motivate physicists and chemists to study monolayers of condensed phase, principally, from viewpoints of the physics of liquid crystals.

Rudnick and Bruinsma argued the non-centrosymmetric orientational structures of monolayers using the elastic theory of nematic liquid crystals, and predicted the shapes of domains of condensed

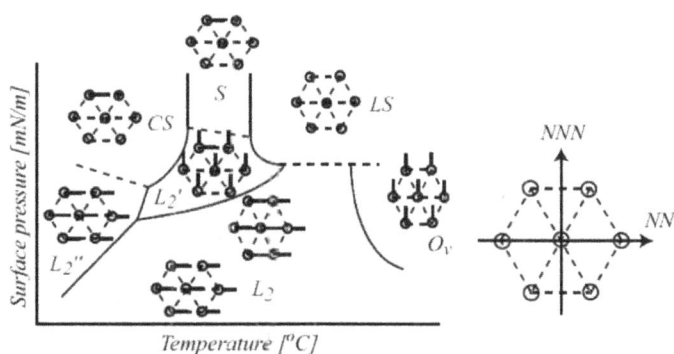

Fig. 4.1:   The mesophases of condensed phase. These phases are characterized by hexagonal alignment of molecular heads and orientational order of chains. Chains are normal to the water surface in the high surface pressure phases, CS, S, and LS, and they are tilted from the water surface normal in the low surface pressure phases, L_2, L_2', L_2'', and Ov. The tilting direction is limited to the nearest neighbor (NN) and the next nearest neighbor (NNN) due to the hexagonal packing of molecular heads. The hexagonal lattice is deformed by the tilting and/or the backbone orientational orders. These orientational and translational orders are shown in the phase diagram by the constituent molecules that are indicated by circles (heads) with rods (chain) on the deformed (or intact) hexagonal lattice. This figure is reproduced from Ref. [1].

phase that coexist with domains of liquid expanded phases in fatty acid monolayers [2]. In other words, their work showed a way to analyze domain shapes in monolayers with consideration of their liquid-crystalline nature. This analytical method is treated in Chapter 5.

The elastic parameters of domains in monolayers have been studied in a similar manner [3], from viewpoints of their liquid-crystalline nature. Fuller's group studied the rheological properties of monolayers in condensed phase [4]. In condensed phases, monolayers are composed of a mosaic of domains, where these domains are in the same thermodynamic phase, but are characterized by the average tilting direction of constituent molecules; the difference of average tilting directions makes a contrast in Brewster angle micrographs. In L_2 phases, the shapes of domains are deformed by applied shear flows, but these shear flows do not couple with the orientations of constituent molecules. In contrast, pure shear flows drive the reorientations of the average tilting directions (tilt azimuth) of constituent molecules in L_2' phases. Experimental optical micrographs such as

BAM images suggested that flow induced reorientations are not initiated uniformly, but proceed by growing stripe regions called "shear bands", where the average tilting directions of constituent molecules are redirected to parallel or antiparallel to the direction of extensions [5]. Maruyama *et al.* proposed that these shear bands are due to the Bingham plasticity of monolayers [4]. Their proposal predicted the dynamics of the width of shear bands roughly in agreement with experiments, though it was not so satisfactory to predict the reorientations of constituent molecules. Indeed, MDC-SHG experiments that can probe polarization of monolayers demonstrated that p-heptyloxy-p'-cyanobiphenyl (7OCB) molecules in monolayers show flow induced reorientations, see Fig. 4.2(b) [5–7]. Optical second harmonic lights were not detected from a 7OCB monolayer at p–s and s–s polarizations in the low surface pressure phase; that monolayer showed the C_∞-symmetry, see region 1 A and B in Fig. 4.2(b). By contrast, in the condensed phase, p–s and s–s second harmonic lights were clearly detected from the 7OCB monolayer; monolayers showed the C_s-symmetry (where the mirror plane was perpendicular to the plane of incidence), at least, in the range of spot size ($\sim 1\,\mathrm{mm}$), see region 2 in Fig. 4.2(b) [6–8]. This MDC-SHG experimental results implied that constituent molecules in 7OCB monolayer show *in-plane* orientational order in the directions of monolayer compressions. The generation of p–s and s–s second harmonic lights were reproducible, and suggested that flows arising from compressions of monolayers annealed the orientations of constituent molecules. The orientational dynamics of monolayers of the condensed phase are discussed in Section 4.3.

One may think flow induced reorientations in monolayers are the 2D analogue of flow induced reorientations in 3D-bulk nematic liquid crystals (where the average tilting directions of monolayers correspond to the average orientations of molecules in bulk nematic liquid crystals). The average orientations of the molecules of nematic liquid crystals are specified by using a unit vector called *directors* and the continuum theories of bulk nematic liquid crystals use director fields to describe their mechanics and dynamics. Ericksen–Leslie theory predicts that, in simple shear flows, the directors of bulk nematic liquid crystals are redirected or show tumbling depending on the values of two viscous constants of materials. Ignès-Mullol and Schwartz showed that the constituent molecules of monolayers show continuous reorientations that are interrupted by discontinuous jumps; this is contradictory to the case of three-dimensional bulk nematic liquid crystals, where their directors show continuous reorientations [9]. They argued that these jumps of reorientations are ascribed to the fact that constituent molecules in L'_2 phases show hexagonal translational order.

As mentioned above, experimental results suggest that monolayers have properties that are analogous to liquid crystals, meanwhile they show properties that are analogous to crystals. That is, amphiphilic molecules in monolayers possess both the lattice-structure property related to the positional ordering of polar heads on the water surface and the liquid crystalline property related to the orientational ordering of chains of long tails of amphiphiles pointing toward air. It is thus instructive to study the transition in the flow of a two-dimensional lattice structure and molecular orientation that are induced by compression, from viewpoints of physics of liquid crystals and solid crystals [7, 10].

In the following sections, we show a way to treat monolayers theoretically, by paying attention to the crystal-like behaviors of monolayers as well as the liquid-crystal-like behaviors of monolayers [7, 10]. To help read this chapter, the physics of general bulk liquid crystals is shortly summarized in Appendix B.

4.2.   Dielectric Polarizations of Monolayers in Condensed Phase: Director Representation

In this section, liquid-crystalline nature of monolayers is further reviewed based on experimental evidence, where dielectric polarizations of monolayers in the condensed phase are represented using *directors*. *Directors* are defined and used in standard textbooks of the Physics of Liquid-crystals [11–14], and we also use them in the same way for monolayers, assuming that *directors* are pointing along ordered long tails of amphiphiles.

Large intensities of second harmonic (SH) lights at the p–s and s–s polarizations are detected from 7OCB (see Fig. 4.2(b)) monolayers in the condensed phase, see Fig. 4.2(c) [7]. Noteworthy that the source of SH lights is the nonlinear polarization of monolayers that is induced by incident laser. Results of Fig. 4.2(c) imply that the orientational structures of monolayers in regions 2 and 3 have the C_s-symmetry, not the C_∞-symmetry, because of their long-range in-plane orientational ordering (namely, orientational ordering with respect to the tilting directions of the long-axes of constituent rod-shaped molecules, see molecular structure of 7OCB in Fig. 4.2(a).

The dielectric physics of monolayers is described in Chapter 2, but it is for the case of monolayers with C_∞-symmetry. To capture an intuitive picture of the structures of ordered monolayers due to tilting, it is convenient to introduce directors **m** that are tilted from

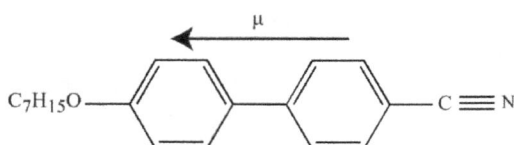

Fig. 4.2(a):   The molecular structure of *p*-heptyloxy-*p'*-cyanobiphenyl (7OCB) molecules. This molecule has a molecular structure that is similar to 8CB molecules, but has one oxygen atom at the connection between their hydrocarbon chains and backbones. 7OCB molecules show isotropic nematic phase transitions at 75°C and nematic-crystal transitions at 54°C.

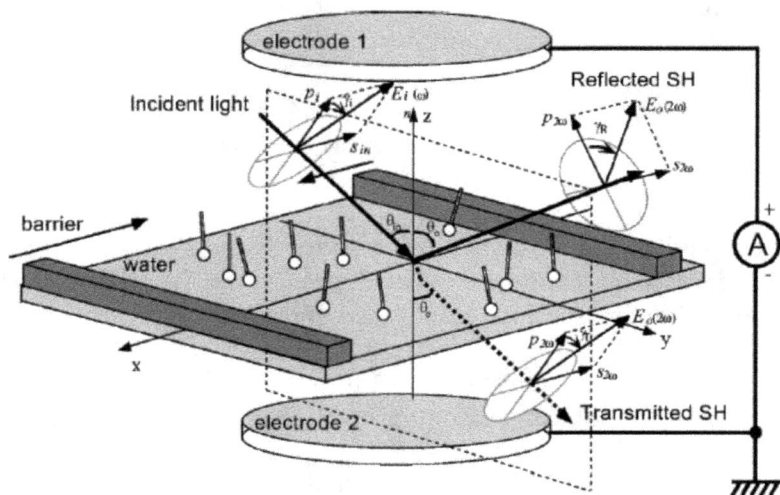

Fig. 4.2(b):   Experimental setup for MDC and SHG measurement. The monolayer is described as the $z = 0$ plane and two moving barriers are described as the two lines $x = +x_0$ and $-x_0$ on the plane at $t = 0$.

Source: Reproduced from Fig. 2 in Ref. [7], *Phys. Rev. E*, 67, 041711, 2003.

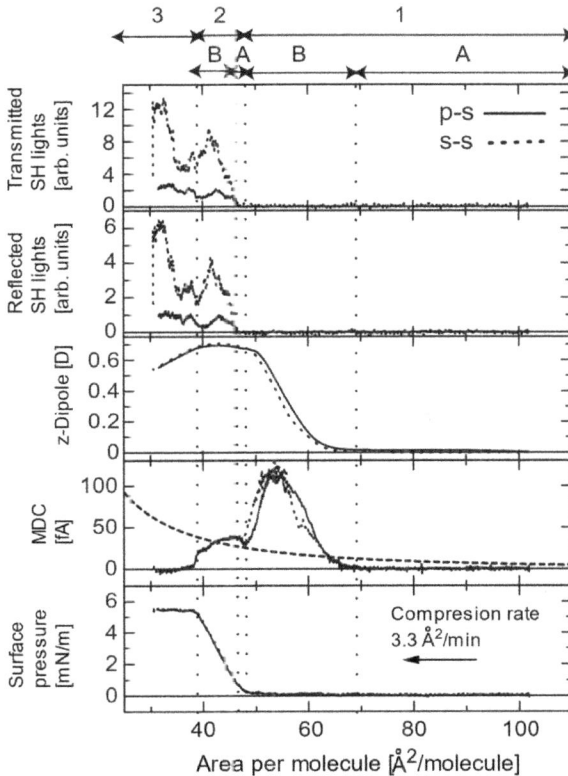

Fig. 4.2(c):   A typical result of MDC-SHG measurement on a 7OCB monolayer in a compression (compression rate is 3.3 $\text{Å}^2/\text{min} \cdot \text{mol}$). The first, second, and third graphs are surface pressure, MDC, and the z-component of dipole moments. We measured the intensities of second harmonic lights in p–s and s–s polarizations. *Source*: Reproduced from Fig. 3 in Ref. [7], *Phys. Rev. E*, 67, 041711, 2003.

the monolayer normal with respect to water surface [7]; this is a natural extension of the case of monolayers with C_∞-symmetry. The idea of using directors **m** can reduce the number of orientational order parameters involved in monolayers, and can provide an intuitive picture of monolayers that are ordered by tilting of the long axis of molecular chains.

The director representation can easily treat the cases that the long-axes of constituent rod-shaped molecules are distributed with C_∞-symmetry around directors **m**. For the cases of monolayers, whose directors $\mathbf{m}$ are normal to air–water interfaces, i.e., $\mathbf{m} = \mathbf{n}$, the orientational structures of these monolayers have C_∞-symmetry. By contrast, the C_s-symmetry is the case that C_∞-symmetry is broken due to the tilting of directors $\mathbf{m}$, i.e., $\mathbf{m} \neq \mathbf{n}$, and constituent molecules show orientational ordering toward the tilting directions of directors.

The extent of orientational ordering of constituent rod-shaped molecules is represented using the orientational order parameters S_n that has the form

$$S_n = \langle P_n(\cos \theta) \rangle, \qquad (4.1)$$

where θ represents tilt angles between the director $\mathbf{m}$ and the long-axes of individual molecules. At a first glance, orientational order parameters S_n for C_s-symmetry, Eq. (4.1), is the same as the orientational order parameters defined for C_∞-symmetry, Eq. (1.3). However, for the C_∞-symmetry, the orientational order parameters S_n are defined by using tilt angles θ of individual molecules from interface normal $\mathbf{n}$, whereas for the C_s-symmetry, these order parameters S_n are defined by using tilt angles θ from the direction of directors $\mathbf{m}$. In other words, Eq. (4.1) is a natural extension of the orientational order parameters for monolayers with C_∞-symmetry to monolayers with C_s-symmetry, where the orientations of constituent molecules are distributed with C_∞-symmetry around directors $\mathbf{m}$.

The orientational distribution functions have the form of θ and ϕ, where θ and ϕ are the tilt angles and tilt azimuth angles of individual molecules with respect to directors $\mathbf{m}$. We thus use the director coordinate system (x_d, y_d, z_d), where the z_d-direction is parallel to the director $\mathbf{m}$ and the x_d- and y_d-directions are perpendicular to the director. The subscripts d indicates the director coordinate system. In general, directors $\mathbf{m}$ are distributed non-uniformly in monolayers and thus the director coordinate is a local coordinate that treats molecules in an area element ΔA that is larger than the molecular length scale and smaller than the length scale of the system.

As the orientations of constituent molecules are distributed with C_∞-symmetry in the director coordinate system, spontaneous, linear, and second-order nonlinear polarizations of monolayers with C_s-symmetry are expressed using the director $\mathbf{m}$ as below:

Dielectric polarizations that are represented in the director coordinate system (x_d, y_d, z_d) are translated to the expressions in the monolayer coordinate system (x, y, z) by using the Euler rotational matrix that has the form

$$R(\phi_d, 0, \theta_d) = \begin{pmatrix} \cos \phi_d \cos \theta_d & -\sin \phi_d & \cos \phi_d \sin \theta_d \\ \sin \phi_d \cos \theta_d & \cos \phi_d & \sin \phi_d \sin \theta_d \\ -\sin \theta_d & 0 & \cos \theta_d \end{pmatrix}, \qquad (4.2)$$

where θ_d and ϕ_d are the tilt angle and tilt azimuth of the director $\mathbf{m}$ in the monolayer coordinate system; θ_d is the angle between the monolayer normal $\mathbf{n}$ and the director $\mathbf{m}$ and ϕ_d is the angle between the x-direction and the projection of the director $\mathbf{m}$ onto the plane of interfaces. The director has the form $\mathbf{m} = (\sin \theta_d \cos \phi_d, \sin \theta_d \sin \phi_d, \cos \theta_d)$ in the monolayer coordinate system.

For the cases of monolayers with C_∞-symmetry, spontaneous polarizations have the form of $\mathbf{P}_0 = N_s \mu_0 S_1 \mathbf{n}$ ($\mathbf{n} = (0, 0, 1)$), meanwhile for the cases of monolayers with C_∞-symmetry that are tilted in the director directions, $\mathbf{m}$, spontaneous polarizations are represented in the form

$$\mathbf{P}_0 = N_s \mu_0 S_1 \mathbf{m}, \qquad (4.3)$$

by using the Euler rotational matrix R. This polarization $\mathbf{P}_0$ corresponds to the spontaneous polarization of monolayers with C_s-symmetry caused by tilting.

Expressions of linear and second-order nonlinear polarizations can be derived in a similar manner by using the Euler rotational matrix R: Linear polarizations of monolayers that have C_∞-symmetry $\chi^{(1)}_{xx} = \chi^{(1)}_{yy} = \chi^{(1)}_{\parallel}$, $\chi^{(1)}_{zz} = \chi^{(1)}_{\perp}$, $\chi^{(-)}_{xy} = -\chi^{(1)}_{yx} = \chi^{(1)}_{ch}$ and other components are zero in the monolayer coordinate system. By contrast, for monolayers that show C_s-symmetry caused by tilting, the components of linear susceptibility tensors have the same as the C_∞-symmetry monolayers in the director coordinate system; the subscripts x, y, and z of $\chi(1)$ should be replaced with x_d, y_d, and z_d. These components of linear susceptibility tensors are translated to the expressions in the monolayer coordinate system by using Euler rotational matrix in the manner as $\chi^{(1)}_{\alpha\beta} = R_{\alpha\mu} R_{\beta\nu} \chi^{(1)}_{\mu\nu}$, where α and β are x, y, and z, and μ and ν are x, y, and z. By using the linear susceptibilities expressed in the monolayer coordinate, linear polarizations generated from mono-layers of C_s-symmetry have the form

$$\mathbf{P}^{(1)} = \chi_{\parallel}^{(1)}\mathbf{E} + (\chi_{\perp}^{(1)} - \chi_{\parallel}^{(1)})(\mathbf{E} \cdot \mathbf{m})\mathbf{m} + \chi_{ch}^{(1)}\mathbf{E} \times \mathbf{m}. \qquad (4.4)$$

Here the components $\chi_{\perp}^{(1)}, \chi_{\parallel}^{(1)}$ and $\chi_{ch}^{(1)}$ have the forms that are shown in Chapter 2 (see Eqs. (2.66)–(2.68)).

With a similar procedure, the expression of second-order nonlinear polarizations generated from monolayers with C_s-symmetry caused by tilting is derived by using the Euler rotational transformation

$$\mathbf{P}^{(2)} = s_{31}E^2\mathbf{m} + (s_{33} - s_{31} - 2s_{15})(\mathbf{E} \cdot \mathbf{m})^2\mathbf{m}$$
$$+ 2s_{15}(\mathbf{E} \cdot \mathbf{m})\mathbf{E} + 2s_{14}(\mathbf{E} \cdot \mathbf{m})(\mathbf{E} \times \mathbf{m}), \qquad (4.5)$$

where the components s_{14}, s_{15}, s_{31}, and s_{33} have the forms that are shown in Chapter 2 for monolayers with the C_∞-symmetry. Indeed, Eqs. (4.3)–(4.5) are derived just by replacing the monolayer normal $\mathbf{n}$ in Eqs. (2.62), (2.68), and (2.75) with the director $\mathbf{m}$, assuming that constituent molecules are distributed with C_∞-symmetry around the director $\mathbf{m}$.

Equation (4.5) is useful for analyzing the large intensities of second harmonic lights generated from 7OCB monolayers in condensed phase in experiments, because it is expressed in a vector form. The intensities of second harmonic lights in s–s and p–s polarizations have the forms

$$I_{p-s} \propto |s_{31} + (s_{33} - s_{31} - 2s_{15})(\sin\theta_i \sin\theta_d \sin\phi_d + \cos\theta_i \cos\phi_d)^2|^2$$
$$\times \sin^2\theta_d \cos^2\phi_d \qquad (4.6)$$

$$I_{s-s} \propto |s_{31} + 2s_{15} + (s_{33} - s_{31} - 2s_{15})\sin^2\theta_d \cos^2\phi_d|^2$$
$$\times \sin^2\theta_d \cos^2\phi_d, \qquad (4.7)$$

where both reflected and transmitted second harmonic lights have the same form because the electric fields of s-polarized lights are both in the x-direction, $\mathbf{s}_r = \mathbf{s}_t = (-1, 0, 0)$, see Figs. 4.2(b) and 4.3. We used the monolayer coordinate, where the z-direction is normal to the monolayer plane and the x-direction is parallel to barrier motions on the water surface, and thus ϕ_d is the angle between the director and the x-axis.

The intensities of s–s second harmonic lights are largest at $\phi_d = 0$ or π, see Eq. (4.7) (for the cases that $(s_{33} - s_{31} - 2s_{15}) \sin^2 \theta_d / (s_{31} + 2s_{15}) \simeq S_3 \sin^2 \theta_d / (3(S_1 - S_3))$ is a positive value, more precisely, a value that is larger than -0.2). This implies that the director of 7OCB monolayers is tilted in condensed phase, $\theta_d \neq 0$, and is directed toward the direction of barrier motion. The intensities of p–s second harmonic lights have a more complex form, see Eq. (4.6), but a similar argument leads to the same conclusion.

Why the director is always directed in parallel or anti-parallel to the barrier movement. This will be argued in Section 4.3, in terms of the effect of flow induced reorientation.

Noteworthy that the use of director representation is powerful for analyzing the observed micrographs of in-plane orientational ordering in the condensed phase of monolayers that are analogous to liquid crystals in smectic C phase. In many cases, smectic C phase is a low temperature phase of smectic A phase, in which constituent rod-shaped molecules show layer structures and the long-axes of these molecules are, on average, normal to the layer. Indeed, experiments have suggested that there are two types of smectic A–C transitions: conventional transitions, directors are tilted from layer normal while the orientational distributions of constituent molecules around the directors are mostly fixed. In the other type of transitions (*de Vries* transitions), the distribution of tilt angles (between the layer normal and individual molecules) does not change significantly but shows orientational ordering with respect to their tilting direction. The director representation shown here is appropriate when the orientational distribution of monolayers in the condensed phase is analogous to that of "conventional" type smectic C liquid crystals. For the cases where monolayers in the condensed phase are analogous to de Vries type, it is more appropriate to use other order parameters that include in-plane orientational ordering, e.g., $\langle \cos \theta \cos \phi \rangle$. Even in the latter case, the director representation is effective as an approximate treatment.

4.3. Flow-Induced Reorientation in Monolayers

In this section, we continue to discuss the liquid-crystalline nature of monolayers, in terms of flow-induced reorientation. In the condensed

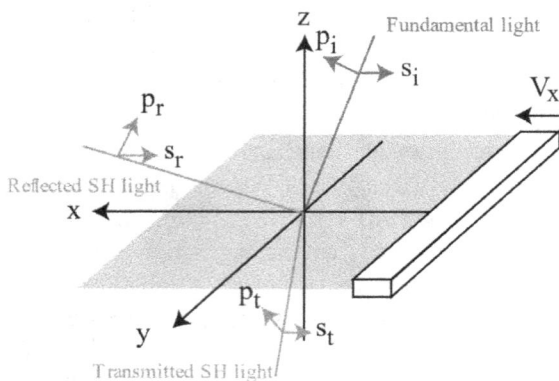

Fig. 4.3:   The monolayer coordinate system, where the z-direction is normal to the water surface and the x-direction is parallel to barrier motions. The y–z plane is the incident plane of second harmonic light measurements and thus the electric fields of s-polarized lights are parallel to the x-direction.

phase, the directors **m** of 7OCB monolayers show reorientation toward the x-direction of Fig. 4.3, during the course of monolayer compression.

This implies that the reorientation of the directors is driven by stresses generated during monolayer compressions. This situation is analogous to flow induced reorientation in bulk nematic liquid crystals, where the reorientations of directors in nematic liquid crystals are driven by viscous stresses due to molecular flows.

The hydrodynamic theory of nematic liquid crystals, Ericksen–Leslie theory, predicts that the asymmetric part of viscous stresses, viscous torques, drive flow induced reorientation in nematic liquid crystals [13, 14]. Employing this idea in the case of two-dimensional monolayers, we can analyze the flow induced reorientation of monolayers. Because of their specific 2D geometry of monolayers, monolayers show anisotropic 2D compressibility in condensed phase; these monolayers are more compressible in the tilting direction of directors than in the perpendicular direction because these monolayers can accommodate decreased area by decreasing the tilt angle of directors. Viscous torques thus drive reorientation toward the direction of compression to accommodate the area decreased by the compression.

4.3.1. *Anisotropic 2D incompressibility*

The hydrophobic chains of rod-shaped molecules are tilted in the nearest neighbor (NN) in L_2 phase that is a low pressure mesophase of condensed phase and in the next nearest neighbor (NNN) in L'_2 and Ov phases that are higher pressure mesophases (see Fig. 4.1). In these phases, the hydrophilic heads of rod-shaped molecules are arranged in hexagonal lattices (hexagonal order) and these hexagonal lattices are deformed either in the NN or NNN direction. Grazing X-ray diffraction experiments suggested that both the tilting of molecular chains and the hindered rotation of these molecules around their long-axes contribute to the lattice deformation.

In the following, we treat a simple case that the hexagonal lattice of molecular heads is deformed by the tilting of molecular chains.

Sirota's theory suggests that molecular area $A(t)$ has a relationship [15]

$$A(t) = \frac{A_u}{\cos \theta_d}, \tag{4.8}$$

where θ_d is the tilt angle of the director, where A_u is the area per molecule of *untilted* phase, see Fig. 4.4; this theory addresses the relationship between lattice deformation and the tilting of directors.

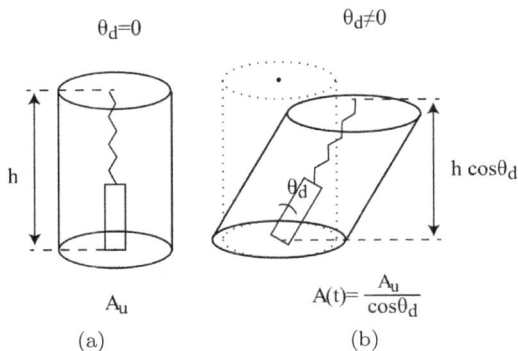

Fig. 4.4: Sirota's cylinder model for rod-shaped molecules in *untilted* phase (a) and *tilted* phase (b). This model assumes that monolayers are incompressible in 3D space and area per molecule is thus elongated toward tilting direction by the factor of $1/\cos \theta_d$.

This implies that the projected area of constituent molecules elongates toward the direction of tilting while their volumes are fixed.

In the theory of elasticity, an elastic material is treated as a continuum that shows Hookean elasticity [14]. In this continuum description, a volume element includes a large number of molecules. Elastic deformations displace a volume element at $\mathbf{a}$ to another position $\mathbf{b}$. The displacements of volume elements are represented by displacement vectors

$$\mathbf{u} = \mathbf{b} - \mathbf{a}. \tag{4.9}$$

It is important to note that both positional vectors $\mathbf{b}$ after an elastic deformation and displacement vectors $\mathbf{u}$ are functions of the original positions $\mathbf{a}$ of the volume elements in materials. The transformation of volume elements $\mathbf{a}, \mathbf{b}$ includes not only elastic deformations but also translations and rotations. In contrast to translations and rotations, elastic deformations change the distance between two points in a material; the extent of elastic deformations is characterized by using strain tensors $\mathbf{u}$ that are defined by an equation

$$\mathbf{db} \cdot \mathbf{db} - \mathbf{da} \cdot \mathbf{da} = 2 \sum_{i=x,y,z} \sum_{j=x,y,z} u_{ij} d\, a_i d\, a_j, \tag{4.10}$$

where u_{ij}, $i, j = x, y$, and z, is the (i, j)-component of strain tensor, and da_i, $i = x, y$, and z, is the i-component of the vector $d\mathbf{a}$. The left-hand side of Eq. (4.10) is the changes in the distance between two points (where a positional vector from one point to the other is $d\mathbf{a}$) by elastic deformation. A straight forward calculation by using Eq. (4.9) leads to the (i, j)-component of strain tensor in the form [16]

$$u_{ij} = \frac{1}{2} \left(\frac{\partial u_j}{\partial a_i} + \frac{\partial u_i}{\partial a_j} \right) + \frac{1}{2} \sum_{k=x,y,z} \frac{\partial u_k}{\partial a_i} \frac{\partial u_k}{\partial a_j}. \tag{4.11}$$

Elastic strains are defined by the *gradient* of displacement vectors; translations and rotations (that are represented by uniform displacement vectors) are eliminated by this gradient. Elastic energy stored in hexagonal lattice is derived by enforcing the symmetry of hexagonal lattices and has the general form [10, 16, 17]

$$f_{\mathrm{ela}} = \frac{c_{11}}{2}(s_1^2 + s_2^2) + c_{12}s_1s_2 + c_{13}(s_1s_3 + s_2s_3)$$

$$+ \frac{c_{44}}{2}(s_4^2 + s_5^2) + \frac{c_{66}}{2}s_6^2, \tag{4.12}$$

with $s_2 = u_{22}$, $s_3 = u_{33}$, $s_4 = u_{23} + u_{32}$, $s_5 = u_{31} + u_{13}$, $s_6 = u_{12} + u_{21}$, and $c_{66} = (c_{11} - c_{12})/2$. We take the relationship, Eq. (4.8), into account in the theory of elasticity. For the case of monolayers that consist of *untilted* molecules, the (average) positions of molecular heads at the nearest neighbor have the form

$$d\mathbf{a}_n = \frac{2d_0}{\sqrt{3}} \left(\cos \frac{n\pi}{3}, \sin \frac{n\pi}{3}, 0 \right)_{\text{hex}}, \tag{4.13}$$

with $n = 0, 1, 2, 3, 4$, and 5 in the hexagonal coordinate system $(x_{\text{hex}}, y_{\text{hex}}, z_{\text{hex}})_{\text{hex}}$, where the z_{hex}-direction is normal to the monolayer and the x_{hex}-direction is parallel to the NN-direction (the y_{hex}-direction is parallel to the NNN-direction), see Fig. 4.5(a). The subscript "hex" indicates that the components of vectors are represented in the hexagonal coordinate system.

We treat the case that molecular chains are tilted in an arbitrary direction $\mathbf{m}_\parallel$, where the angle between $\mathbf{m}_\parallel$ and the NN-direction is ϕ_M ($\phi_M = 0$ is the NN-direction and $\pi/2$ is the NNN-direction). The hexagonal lattice of molecular heads is deformed in the tilting direction of molecular chains according to Eq. (4.8), see Fig. 4.5. In this case, it is useful to introduce the tilt coordinate system $(x_{\text{til}}, y_{\text{til}}, z_{\text{til}})$,

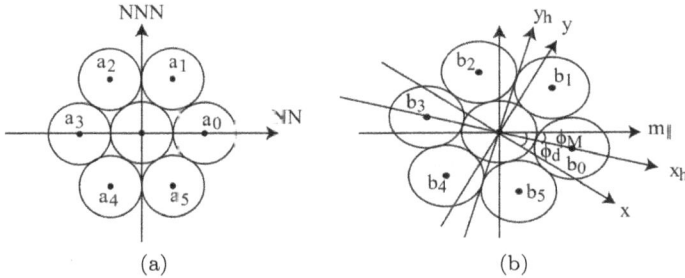

(a)   (b)

Fig. 4.5: The hydrophilic heads of constituent molecules are aligned in a hexagonal lattice in the monolayers of condensed phase. For the case of monolayers, where the hydrophobic chains of constituent molecules are not tilted, the hexagonal lattice of their hydrophilic heads does not show deformation (a). We here treat the cases of monolayers, where the hexagonal lattice of the hydrophilic heads of constituent molecules shows elastic deformations toward the direction $\mathbf{m}_\parallel$ that their hydrophobic chains are tilted. The extent of these elastic deformations given by Sirota's theory, Eq. (4.8). ϕ_M is the angle between the nearest neighbor (NN) direction and $m\perp$ and ϕ_d is the angle between $\mathbf{m}_\parallel$ and the direction of monolayer compression (the x-direction in monolayer coordinate) (b).

where the x_{til}-direction is parallel to $\mathbf{m}_{\parallel}$ and the z_{til}-direction is normal to the monolayer. In this coordinate system, the positional vectors $d\mathbf{a}_n$ have the form

$$d\mathbf{a}_n = \frac{2d_0}{\sqrt{3}}\left(\cos\left(\frac{n\pi}{3} - \phi_M\right), \sin\left(\frac{n\pi}{3} - \phi_M\right), 0\right)_{\text{tilt}}, \qquad (4.14)$$

with $n = 0, 1, 2, 3, 4$, and 5, see also Fig. 4.6. Equation (4.8) suggests that the hexagonal lattice of molecular heads is deformed by the tilting of molecular chains by $1/\cos\theta_d$, see also Fig. 4.5; the tilting of molecular chains displaces the molecular heads at the nearest neighbor to new positions

$$d\mathbf{b}_n = \frac{2d_0}{\sqrt{3}}\left(\frac{\cos\left(\frac{n\pi}{3} - \phi_M\right)}{\cos\theta_d}, \sin\left(\frac{n\pi}{3} - \phi_M\right), 0\right)_m, \qquad (4.15)$$

with $n = 0, 1, 2, 3, 4$, and 5, in the tilt coordinate system.

Substituting Eqs. (4.13)–(4.15) into Eq. (4.10) leads to the forms of in-plane components of strain tensor

$$s_1^{\text{hex}} = u_{11}^{\text{hex}} = \frac{1}{2}\tan^2\theta_d\cos^2\phi_M, \qquad (4.16)$$

$$s_2^{\text{hex}} = u_{22}^{\text{hex}} = \frac{1}{2}\tan^2\theta_d\sin^2\phi_M, \qquad (4.17)$$

$$s_6^{\text{hex}} = u_{12}^{\text{hex}} + u_{21}^{\text{hex}} = \tan^2\theta_d\sin\phi_M\cos\phi_M, \qquad (4.18)$$

in the hexagonal coordinate system.

In the monolayer coordinate (x, y, z), where the z-direction is normal to the monolayer and the x-direction is parallel to the compression direction, the components of strain tensor have the forms

$$s_1 = u_{11} = \frac{1}{2}\tan^2\theta_d\cos^2\phi_d, \qquad (4.19)$$

$$s_2 = u_{22} = \frac{1}{2}\tan^2\theta_d\sin^2\phi_d, \qquad (4.20)$$

$$s_6 = u_{12} + u_{21} = \tan^2\theta_d\sin\phi_d\cos\phi_d, \qquad (4.21)$$

where ϕ_d is the azimuthal angle of director (the angle between the x-direction and the in-plane projection $\mathbf{m}_{\parallel}$ of director).

Substituting Eqs. (4.19)–(4.21) into Eq. (4.12) leads to elastic energy in the form

$$f_{\text{ela}} = \frac{1}{8} c_{11} \tan^4 \vartheta_d - \frac{1}{4} (\Pi_{11}^0 + \Pi_{22}^0) \tan^2 \theta_d$$

$$- \frac{1}{4} (\Pi_{11}^0 - \Pi_{22}^0) \tan^2 \theta_d \cos 2\phi_d, \qquad (4.22)$$

where Π_{11}^0 and Π_{22}^0 are applied stresses. The second and third terms are the work done by these applied stresses. Equation (4.22) is independent of ϕ_M; the fact that molecular heads form hexagonal lattice do not play any role in the reorientations of directors. The last term of Eq. (4.22) depends on the in-plane orientation ϕ_d of directors and is proportional to $\Pi_{11}^0 - \Pi_{22}^0$. This implies that directors show reorientation toward the direction that the largest stress is applied because of the anisotropic 2D compressibility.

4.3.2. *Flow-induced reorientation of monolayers* [6, 10]

We here extend the hydrodynamic theory of nematic liquid crystals to the case of the flow induced reorientation of monolayers. Because of the anisotropic 2D compressibility, the elastic energy, Eq. (4.22), also contributes to torque balance equations, see Appendix B, where directors $\mathbf{m}$ are pointing in the direction along long tails of amphiphiles pointing toward air that is orientationally ordered. Torque balance equation thus has the form

$$I \frac{d}{dt} \left(\mathbf{m} \times \frac{d\mathbf{m}}{dt} \right) = \Gamma_{\text{Fra}} + \Gamma_{\text{vis}} + \Gamma_{\text{ela}}, \qquad (4.23)$$

where Γ_{ela} is torque arising from anisotropic 2D compressibility and has the form, see also Eq. (4.22),

$$\Gamma_{\text{ela}} = \mathbf{m} \times \left(\frac{\delta f_{\text{ela}}}{\delta \mathbf{m}} \right). \qquad (4.24)$$

Elastic stresses σ^{ela} arising from deformations of hexatic stresses contribute to Navier–Stokes equation that has the form

$$\rho \frac{d}{dt} v_\alpha = \partial_\beta (-\Pi_{\beta\alpha} + \sigma_{\beta\alpha}^{\text{vis}} + \sigma_{\beta\alpha}^{\text{ela}}), \qquad (4.25)$$

where α and β are x, y, and z. Because of the fact that 2D compressibility is anisotropic [1], surface pressure Π_{ij} is anisotropic. The orientational distribution of director $\mathbf{m}$, surface pressure Π, and velocity fields $\mathbf{v}$ are determined by solving Eqs. (4.23) and (4.25) together with 2D analogue of the equation of mass conservation, see Appendix B, where ρ is the area density of constituent molecules (the inverse of area per molecule $A(t)$) of monolayers and $\mathbf{r}$ is 2D positional vectors in monolayer planes.

In a compression of monolayers with a constant barrier speed, area per molecule $A(t)$ has the form

$$A(t) = A_0, \qquad (4.26)$$

where A_0 is area per molecule at $t = 0$ and V_A is compression rate that is the decrease of area per molecule in a unit time. Using Eq. (4.26) leads to the form of velocity fields $\mathbf{v} = (v_x, 0, 0)$ with

$$v_x = -\frac{V_A}{A(t)}x. \qquad (4.27)$$

Note that we treat the case of monolayers, where 2D anisotropic compressibility accommodates all area decreases by a monolayer compression, and thus these monolayers do not show flows in the y-direction that is perpendicular to the direction of monolayer compressions.

We here use the monolayer coordinate, where the z-direction is normal of the monolayer and the x-direction is parallel to the direction of monolayer compression, see Fig. 4.3. The xx-component d_{xx} of the shear rate tensor is $-V_A/A(t)$ and the other components of this tensor are zero. The vorticity $\omega \sim \left(\equiv \frac{1}{2}\nabla \times v\right)$ is zero. The director $\mathbf{m}$ has the form $\mathbf{m} = (\sin\theta_d \cos\phi_d, \sin\theta_d \sin\phi_d, \cos\theta_d)$ in the monolayer coordinate, see Fig. 4.4. The rate $\mathbf{M}$ of director reorientation (relative to vortex flows) is calculated.

We here treat the case that director reorientation with respect to tilt angle is much slower than director reorientation with respect to tilt azimuth, $\dot\phi_d \gg \dot\theta_d$ in this case, the time derivatives of tilt angle are negligible, $\dot\theta_d \sim 0$. Substituting the form of $\mathbf{M}$, $\mathbf{m}$, and $\mathbf{d}$ into Eq. (B.38) in Appendix B leads to the form of viscous stresses.

$$\sigma_{xx}^{\text{vis}} = \mu_1 d_{xx}\sin^4\theta_d \cos^4\phi_d - (\mu_2 + \mu_3)\dot\phi_d \sin^2\theta_d \cos^2\phi_d + \mu_3 d_{xx}$$
$$+ (\mu_5 + \mu_6)d_{xx}\sin^2\theta_d \cos^2\phi_d, \qquad (4.28)$$

$$\sigma_{xy}^{\text{vis}} = \mu_1 d_{xx} \sin^4 \theta_d \cos^3 \phi_d \sin \phi_d$$

$$+ \dot{\phi}_d \sin^2 \theta_d (\mu_2 \cos^2 \phi_d - \mu_3 \sin^2 \phi_d)$$

$$+ \mu_6 d_{xx} \sin^2 \theta_d \sin \phi_d \cos \phi_d, \tag{4.29}$$

$$\sigma_{yx}^{\text{vis}} = \mu_1 d_{xx} \sin^4 \theta_d \cos^3 \phi_d \sin \phi_d$$

$$+ \dot{\phi}_d \sin^2 \theta_d (\mu_3 \cos^2 \phi_d - \mu_2 \sin^2 \phi_d)$$

$$+ \mu_5 d_{xx} \sin^2 \theta_d \sin \phi_d \cos \phi_d, \tag{4.30}$$

$$\sigma_{yy}^{\text{vis}} = \mu_1 d_{xx} \sin^4 \theta_d \sin^2 \phi_d \cos^2 \phi_d$$

$$+ (\mu_2 + \mu_3)\dot{\phi}_d \sin^2 \theta_d \sin \phi_d \cos \phi_d. \tag{4.31}$$

The viscous stresses are not symmetric tensors and thus generate viscous torques, where their z-components, $\Gamma_z^{\text{vis}} = -\sigma_{xy}^{\text{vis}} + \sigma_{yx}^{\text{vis}}$, have the form

$$\Gamma_z^{\text{vis}} = -\sin^2 \theta_d [(\mu_2 - \mu_3)\dot{\phi}_d - (\mu_5 - \mu_6)d_{xx} \sin \phi_d \cos \phi_d]. \tag{4.32}$$

The z-components of torques arising from 2D anisotropic compressibility have the form

$$\Gamma_z^{\text{ela}} = \frac{1}{2}(\Pi_{11}^0 - \Pi_{22}^0) \tan^2 \theta_d \sin 2\phi_d. \tag{4.33}$$

For simplicity, we here treat the case of directors that are uniformly distributed, and torques Γ^{Fra} arising from Frank elastic energy is zero. Because inertia terms are very small in experiments (low Reynolds number limit), torque balance equation (Eq. (4.13)) leads to an equation of director reorientation

$$\frac{d\phi_d}{dt} = F(t)[V_A - V_A^*] \sin 2\phi_d, \tag{4.34}$$

with

$$F(t) = -\frac{1}{2A(t)} \frac{\mu_6 - \mu_5}{\mu_3 - \mu_2}, \tag{4.35}$$

$$V_A^* = -\frac{\Pi_{11}^0 - \Pi_{22}^0}{\cos^2 \theta_d} \frac{A(t)}{\mu_6 - \mu_5}. \tag{4.36}$$

Equation (4.34) suggests that stable director orientation is $\phi_d = \pm\pi/2$ for $F(t)(V_A - V_A^*) > 0$, and $\phi_d = 0$ and π for $F(t)(V_A - V_A^*) < 0$.

Table 4.1:   Ericksen–Leslie coefficients of nematic liquid crystals: MBBA, 5CB, 5OCB, 8OCB. Nematic–isotropic (N–I) transitions of MBBA, 5CB, 5OCB, and 8OCB are at 45°C, 35°C, 68°C, 80°C, respectively.

$[10^{-3}\text{Pa} \cdot \text{s}]$	μ_1	μ_2	μ_3	μ_4	μ_5	μ_6	Refs.
MBBA	6.5	-77.5	-1.2	83.2	46.3	-34.4	[13]
5CB (23°C)	-11	-102	-5	74	288	-23	
5OCB (50°C)		-27	~ 0	46	20	-14	[19]
8OCB (60°C)		-30	~ 0	30	20	-10	[19]

For typical values of the Ericksen–Leslie constants of nematic liquid crystals, $F(t)$ is a positive function (see Table 4.1). V_A^* is the threshold compression rate that determines the stable director orientation.

We here estimate the Ericksen–Leslie constants of mono-layers as the values of the Ericksen–Leslie constants of the bulk nematic liquid crystals multiplied by the thickness of monolayers and the difference $\Pi_{11}^0 - \Pi_{22}^0$ between surface pressures as $\sim 1\,\text{mN/m}$. In this case, the threshold value V_A^* of compression rate is estimated to be $3 \times 10^9\,\text{Å}^2/\text{min}$ which is much larger than the compression rate used in our experiments, $V_A - V_A^* < 0$. This implies that the stable director orientation is toward $\phi_d = 0$ and/or π. These results agree with the results of SHG measurements, see also Section 4.3. The very large values of the threshold compression rate V_A^* may be because we used the Ericksen–Leslie constants of bulk nematic liquid crystals to estimate the values of monolayers in the condensed phase that may be more viscous and used the absolute values of surface pressures to estimate the difference $\Pi_{11}^0 - \Pi_{22}^0$ of surface pressures (between the x-and y-directions); more experiments will be necessary to quantitatively check this argument. Indeed, Eq. (4.34) is explicitly solved in the form

$$\tan \phi_d(t) = \tan \phi_d(0) \exp\left[2 \int_0^t ds\, F(s)(V_A - V_A^*(s))\right]. \qquad (4.37)$$

Noteworthy that by introducing the local lattice elasticity into the pure Ericksen–Leslie EL theory which only considers the deformation of the director, a more general theory can be developed for further understanding the behaviors of hexatic liquid crystal under flow [10]. In the cases of amphiphilic monolayers in tilted phases L_2 and $L_{2'}$, the exact solutions of the new EL equation in two types of flow, pure extension and simple shear, explain well most of the features of flow-induced tilt azimuth orientation observed by Fuller's group [4] and Schwartz's group [9]. In particular, the "shear band" domain generated by flow discovered by the former is proved theoretically as the result of two-dimensional Wulf construction [17] in L_2 and $L_{2'}$ phases [10].

4.4. Flexoelectric Effect of Monolayers [5]

The generation of large intensities of second harmonic lights in s–s and p–s polarizations is because of the fact that monolayer compression drives flow induced director reorientation. Interestingly, in this phase, MDCs decrease with decreasing area per molecule by compression, see region 2 in Fig. 4.2(b). Differentiating Eq. (4.3) with respect to t at molecular area A, the form of MDCs is derived

$$I = \frac{\mu_0 B V_A}{L} \left(\frac{S_1 \cos \theta_d}{A^2} - \frac{1}{A} \frac{d}{dA} (S_1 \cos \theta_d) \right). \tag{4.38}$$

The broken curve in the second graph from the bottom, Fig. 4.2 is the first term of Eq. (4.38) that is calculated from the value of $\mu_0 S_1 \cos \theta_d$ measured in the MDC experiments, see the third graph from the bottom, Fig. 4.2(c). The first term of Eq. (4.38) predicts that MDCs increase monotonously with decreasing molecular area A, but experimental MDCs are smaller than the estimated values. One may think that the second term of Eq. (4.38) is negative, and constituent molecules increase their tilt angles with monolayer compression, though this is unusual. On the other hand, it will be more acceptable to think that decreasing molecular-area A suppresses the tilting of constituent molecules, but it does not lead to a decrease in MDCs by compression. These puzzling experimental MDC results suggest the possibility of the appearance of other types of static dielectric polarizations. The most probable one is due to *flexoelectric effects* predicted by R. B. Meyer [18].

For the cases of bulk nematic liquid crystals, dielectric polarizations arising from flexoelectric effect have the form

$$\mathbf{P}_{flx} = f_s (\nabla \cdot \mathbf{m})\mathbf{m} + f_b \mathbf{m} \times \nabla \times \mathbf{m}. \tag{4.39}$$

This dielectric polarization is generated due to splay deformations or bend deformations, where f_s and f_b are called flexoelectric constants with respect to splay and bend deformations, respectively. In smectic C phase, liquid crystals stabilize layer structures and their directors are tilted from the layer normal. As mentioned in earlier sections, a monolayer in condensed phase is analogous to a layer of liquid crystals in smectic C phase. We thus phenomenologically treat the 2D geometry and non-centrosymmetric orientational structures of monolayers in an extension of the flexoelectric effects of smectic C liquid crystals [5, 20].

We write director $\mathbf{m}$ in the form $\mathbf{m} = (\sin\theta_d \cos\phi_d, \sin\theta_d \sin\phi_d, \cos\theta_d)$, where θ_d is the tilt angle of director and ϕ_d is its azimuthal angle, and treat the case of monolayers, where the tilt angles of their directors are uniform and the azimuthal angle of their directors may be distributed non-uniformly. In this case, Eq. (4.39) is rewritten in a more convenient form

$$\mathbf{P}_{\text{flx}} = f_s \sin^2\theta_d(\nabla_{\parallel} \cdot \mathbf{c})\mathbf{c} + f_s \sin\theta_d \cos\theta_d(\nabla_{\parallel} \cdot \mathbf{c})\mathbf{n}$$
$$- f_b \sin^2\theta_d(\nabla_{\parallel} \times \mathbf{c})(\mathbf{n} \times \mathbf{c}), \tag{4.40}$$

where $\mathbf{c}$ is so-called in-plane directors that are unit vectors to the projections of directors $\mathbf{m}$ onto monolayer planes, $\mathbf{c} = (\cos\phi_d, \sin\phi_d, 0)$ in the monolayer coordinate system. $\nabla_{\parallel}$ is 2D gradient along the monolayer plane and thus

$$\nabla_{\parallel} \times \mathbf{c} = \frac{\partial c_y}{\partial x} - \frac{\partial c_x}{\partial y}, \tag{4.41}$$

is a scalar quantity. $\mathbf{n}$ is the monolayer normal. Equation (4.40) suggests that monolayers generate dielectric polarizations in the monolayer normal $\mathbf{n}$ when they show in-plane splay deformations.

The Frank elastic energy of non-centrosymmetric orientational structures of monolayers is described as

$$f_{\text{Fra}} = \frac{1}{2}k_s(\nabla \cdot \mathbf{m} - s_0)^2 + \frac{1}{2}k_t(\mathbf{m} \cdot \nabla \times \mathbf{m} - t_0)^2$$
$$+ \frac{1}{2}k_b(\mathbf{m} \times \nabla \times \mathbf{m})^2, \tag{4.42}$$

where s_0 is the so-called spontaneous splay and is not zero only for the cases of materials that have non-centrosymmetric orientational structures. This implies that the in-plane directors of monolayers spontaneously show splay deformations in the state of minimal free energy when there are no applied fields that anneal these orientational deformations; this will generate dielectric polarizations due to flexoelectric effects (see Eq. (4.40)).

In contrast to spontaneous polarizations, flexoelectric polarizations are suppressed when the splay deformations of directors are annealed by external fields. e.g., by flow induced reorientation, Indeed, monolayers show flow induced reorientation in the condensed phase, where MDCs decreases with decreasing area per molecule (see Fig. 4.2(b)). This implies that annealing of spontaneous splay deformations suppresses flexoelectric polarizations generated from monolayers and this is the most probable mechanism behind the puzzling MDC experimental results.

Maxwell displacement currents that are generated by flexoelectric polarizations have the form

$$I_{\mathbf{flx}} = \frac{1}{L}\frac{d}{dt}\int_B dS(f_3\nabla_{\|} \cdot c), \tag{4.43}$$

with $f_3 = f_s\sin\theta_d\cos\theta_d$, where B is the area of electrodes and L is the distance between an electrode supported at the air and water surface (see Fig. 4.2(b)). For the case of MDC measurements during monolayer compressions, the time derivative of Eq. (4.43) is thus evaluated as

$$\lim_{\Delta t\to 0}\frac{\int_B dS f_3\nabla_{\|} \cdot c(r + vt) - \int_B dS f_3\nabla_{\|} \cdot c(r)}{\Delta t}$$

$$+ \lim_{\Delta t\to 0}\frac{\int_{B+\Delta S} dS f_3\nabla_{\|} \cdot c(r) - \int_B dS f_3\nabla_{\|} \cdot c(r)}{\Delta t}$$

$$+ \int_B dS f_3\frac{\partial}{\partial t}\nabla_{\|} \cdot c(r), \tag{4.44}$$

where ΔS is the area of molecules that will flow toward the region under the electrode for MDC measurements in Δt (see Fig. 4.6). The first and third terms result from the Storks derivative of $\nabla_{\|} \cdot c$ and the second term results from the fact that constituent molecules are accumulated under the electrode by compressing the monolayer. Molecules in the region of ΔS accumulate under the electrode in Δt (see Fig. 4.6). Compressing a monolayer induces velocity fields and

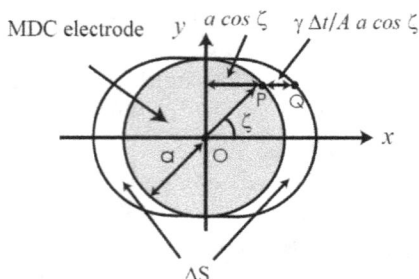

Fig. 4.6:   Compressing monolayers accumulates molecules under MDC electrodes and MDCs generated by this effect are shown in the second term of Eq. (4.44). ΔS in Eq. (4.44) is the white region around MDC electrode (the gray circle) that has the thickness $V_A \Delta t\, A\, a \cos \zeta$, where a is the radius of MDC electrode, ζ is the angle between the x-direction (parallel to the direction of compression) and the position P at the rim of the electrode.

Source: Reproduced from Ref. [5], *J. Chem. Phys.*, 122, 164703, 2005.

thus the width of the region of ΔS is $v_x \Delta t\, V_A a \cos \zeta \Delta t / A$, where a is the radius of the electrode and ζ is the angle between the x-direction (that is parallel to compressing direction) and positions at the rim of the electrode. The third term of Eq. (4.44) thus has the form

$$\int_{\Delta S} dS f_3 \nabla_{\parallel} \cdot \boldsymbol{c}(\boldsymbol{r}) = \frac{V_A}{A} \Delta t \oint_B ds x \nabla_{\parallel} \cdot \boldsymbol{c}(\boldsymbol{r}). \tag{4.45}$$

Substituting Eq. (4.45) into Eq. (4.44) and a few calculations lead to MDCs arising from flexoelectric polarizations in the form

$$I_{\text{flx}} = \frac{B V_A}{L} \frac{\xi}{A} + I_{\text{reo}}, \tag{4.46}$$

with

$$\xi = \frac{f_3}{B} \int_B dS \nabla_{\parallel} \cdot \boldsymbol{c}. \tag{4.47}$$

$$I_{\text{reo}} = \frac{1}{L} \int_B dS f_3 \frac{\partial}{\partial t} \nabla_{\parallel} \cdot \boldsymbol{c}(\boldsymbol{r}). \tag{4.48}$$

The first term is because of the fact that compressing monolayers increases director deformations and the second term is due to the reorientation of directors; flow induced reorientation contributes to MDCs via I_{reo}. For the cases that I_{reo} is zero, the sum of MDCs arising from spontaneous polarizations, Eq. (4.38), and those arising from flexoelectric polarizations, Eq. (4.47), does not predict concave functions as MDCs that were measured in the MDC experiments, see Fig. 4.2(c), even when ξ is a negative value (Note that we here implicitly assume that ξ does not depend on area per molecule). When average flexoelectric polarizations ξ under the electrode are positive, flow induced director reorientation decreases flexoelectric polarizations and thus MDCs decrease $I_{reo} < 0$, Eq. (4.48). For the case that flexoelectric polarizations are completely suppressed by flow induced reorientations, the value of ξ is estimated to be 1.86×10^{-13} C/m from the experimental results, see Fig. 4.2(c). Flexoelectric constants of monolayers are estimated by the typical order of magnitudes 0.01 C/m of flexoelectric constants of nematic liquid crystals multiplied by the thickness of monolayers 1 nm. Initial splay deformations, $\nabla_\| \cdot c$, are estimated to be 1.86×10^{-2} m^{-1}; very small splay deformations are enough to generate MDCs. This implies that the decrease of MDC with decreasing area per molecule is possibly because of the flow-induced director reorientation suppressed flexoelectric polarizations.

4.5.   Summary

Many experimental results suggest that monolayers show behaviors that are analogous to liquid crystals, particularly, in the condensed phase. The details of the dynamics of the orientations of monolayers are indeed different from the dynamics of the orientations of bulk nematic liquid crystals. This is possibly due to the fact that molecular heads are packed in hexagonal lattices in the condensed phase, but they are on the water surface. Nevertheless, treating monolayers as 2D liquid crystals that show polar *in-plane* orientational order accompanied by a flow of hexatic lattice structure can capture many aspects of the orientational dynamics of monolayers in the condensed phase. Continuum theories treat bulk nematic liquid crystals as a field of directors: The directors of bulk nematic liquid crystals are easily deformed to non-uniform distributions even by small perturbations because their orientational ordering arises from the *cooperativity* of constituent molecules. The translational degrees of freedom of constituent molecules are coupled to their orientational degrees of freedom; applied shear flows induce director reorientations (and, conversely, director reorientations induce shear flows) due to viscous torques that are asymmetric components of viscous stresses.

MDC-SHG experiments show that 7OCB monolayers generate large intensities of second harmonic lights at the *s–s* and *p–s* polarizations in condensed phase; this implies that these monolayers show *in-plane* orientational order. In an analogy with the description of smectic C phase of liquid crystal physics, *directors* can be employed. This analytical approach shows that the large intensities of second harmonic lights in the *s–s* and *p–s* polarizations are generated because the reorientations of directors are in the directions *parallel* or *anti-parallel* to the barrier motions. This is because monolayer flows arising from barrier motions induced reorientations of directors in an analogous manner to the flow induced director reorientations of bulk nematic liquid crystals. Considering 2D anisotropic compressibility in an extension of the hydrodynamic theory of nematic liquid crystals, it is shown the directors of monolayers in condensed phase show reorientation toward the direction of monolayer compression.

Maxwell displacement currents (MDC) decreases with decreasing area per molecule, but this is not reasonably predicted by the dynamics of spontaneous polarizations. Consideration of the liquid crystal-like nature of monolayers in condensed phase, generation of dielectric polarizations arising from flexoelectric effects was suggested; the in-plane directors of monolayers spontaneously show splay deformations and thus generate dielectric polarizations due to flexoelectric effects at the state of minimal free energy. Applied shear flows induce flow induced reorientations and thus anneal these splay deformations, and this suppresses the dielectric polarizations and lead to the decrease of MDC.

In summary, monolayers show *in-plane* orientational structures that are easily deformed by small perturbations. This is analogous to nematic liquid crystals. Monolayers show polar *in-plane* orientational orders with flow of molecular heads of hexatic lattice structure; this is in contrast to the situation of nematic liquid crystals that show axial orientational order.

References

[1] V. M. Kaganer, H. Möhwald, and P. Dutta, Structure and phase transitions in Langmuir monolayers, *Rev. Mod. Phys.*, 71, 779–819 (1999).

[2] J. Rudnick and R. Bruinsma, Shape of domains in two-dimensional systems: Virtual singularities and a generalized Wulff construction, *Phys. Rev. Letts.*, 74, 2491–2494 (1995).

[3] S. Rivière and J. Meunier, Anisotropy of the line tension and bulk elasticity in two-dimensional drops of a mesophase, *Phys. Rev. Letts.*, 74, 2495–2498 (1995).

[4] T. Maruyama, G. Fuller, C. Frank, and C. Robertson, Flow-Induced Molecular orientation of a Langmuir film, *Science*, 274, 233–235 (1996).

[5] T. Yamamoto, D. Taguchi, T. Manaka, and M. Iwamoto, Detection of flexo-electric effect from 4-heptyloxy-4′cyanobiphenyl monolayers at an air–water interface by Maxwell displacement current and optical second harmonic generation, *J. Chem. Phys.*, 122, 164703 (2005).

[6] A. Tojima, *Characterization of Dielectric Polarizations Generated from Monolayers by Using Maxwell Displacement Current and Optical Second Har- Monic Generation*, Ph.D. thesis, Tokyo Institute of Technology (Japanese).

[7] M. Iwamoto, A. Tojima, T. Manaka, and Z. C. Ou-Yang, Compression- shear-induced tilt azimuth orientation of amphiphilic monolayers at the air-water interface: A $C_\infty \to C_{2v}$ transition in the flow of a two-dimensional hexatic structure, *Phys. Rev. E*, 67, 041711 (2003).

[8] T. Yamamoto, A. Tojima, T. Manaka, and M. Iwamoto, Detection of the reversible flow behavior of 4-heptyloxy-4′-cyanobiphenyl monolayer at air–water interface by compression and expansion with Maxwell displacement current and optical second harmonic generation, *Chem. Phys. Letts.*, 378, 428–433 (2003).

[9] J. Ignès-Mullol and D. K. Schwartz, Shear-induced molecular precession in a hexatic Langmuir monolayers, *Nature*, 410, 348–351 (2001).

[10] M. Iwamoto and Z.-C. Ou-Yang, Flow-induced molecular orientation of amphiphile monolayers: Incorporation of hexatic elasticity into Ericksen–Leslie theory, *Phys. Rev. E*, 72(2), 21704 (2005).

[11] Liquid crystal handbook editorial committee, *Liquid Crystal Handbook*, Maruzen, 2000 (Japanese).

[12] T. Geelhaar, K. Griesar, and B. Reckmann, 125 years of liquid crystals — A Scientific revolution in the home, *Angew. Chem. Int. Ed.*, 52, 8798–8809 (2013).

[13] P. G. de Gennes and J. Prost, *The Physics of Liquid Crystals*, Oxford University Press, New York, 1993.

[14] S. Chandrasekhar, *Liquid Crystals*, 2nd ed., Cambridge University Press, New York, 1992.

[15] E. B. Sirota, Remarks concerning the relation between rotator phases of bulk n-alkanes and those of Langmuir monolayers of alkyl-chain surfactant on water, *Langmuir*, 13(14), 3849–3859 (1997).

[16] J. F. Nye, *Physical Properties of Crystals*, Oxford University Press, Oxford, 1957.

[17] G. Wulff, XXV. Zur frage der geschwindigkeit des wachsthums und der auflösung der krystallflächen, Z. *Kristallogr.*, 34, 449 1901.

[18] R. B. Meyer, Piezoelectric effects in liquid crystals, *Phys. Rev. Letts.*, 22, 918 (1969).

[19] A. G. Chmielewski, Viscosity coefficients of some nematic liquid crystals, *Mol. Cryst. Liq. Cryst.*, 132, 339 (1986).

[20] T. Yamamoto, *Dielectric and Geometrical Properties of Dipolar Monolayers in Liquid Crystal Phase*, Ph.D. thesis, Tokyo Institute of Technology, Tokyo.

Chapter 5

Differential Geometry Method for Analyzing Domain Shapes

This chapter describes the mathematical approach for analyzing the shapes of domains in monolayers. The mathematical method "differential geometry" is introduced to describe 1D curves on 2D flat surfaces. The free energy of domains in monolayers is introduced based on the theory by McConnell and Moy [1], where the shapes of domains are determined by the competition between line tensions and electrostatic energy arising from interactions between electric dipoles. A mathematical method to treat electrostatic dipolar energy caused from dipole–dipole interactions is shown by using the Frenet–Serret theorem [2], and shape equations describing domain shapes are derived in an approximate form [3]. A variety of domains of shapes, e.g., circular shapes of solid domain surrounded by fluid phase, Clover-leaf shapes (CLS), Kidney- and boojum-like domains that abound in a lipid monolayer, deformed shapes of their transitions to torus, D form, S form, serpentine manner form, m-sided quasi-polygon form, and others can be displayed by solving the shape equations derived. Chapter 6 is a companion chapter, where we further focus on the physics of the shapes of domains in monolayers by deriving the formula satisfying minimum free energy condition, in association with electrostatic Maxwell stress.

5.1. Differential Geometry of 2D Flat Surfaces

The mathematical method of "Differential geometry" [2] is useful for analyzing the shapes of soft matters, including monolayers [3]. In a macroscopic description, where the thickness of the interfaces between domains is negligible, these interfaces are treated as 1D curves that are embedded in 2D flat surfaces. In differential geometry, a 1D curve is treated by a positional vector $\tilde{\mathbf{r}}(u) = (\tilde{x}(u), \tilde{y}(u))$ that is a function of a parameter u (see also Fig. 5.1(a)).

Elementally calculus suggests that the contour length s of this curve has the form

$$ds = \sqrt{\left(\frac{d\tilde{x}}{du}\right)^2 + \left(\frac{d\tilde{y}}{du}\right)^2}\, du$$

$$= \left|\left(\frac{d}{du}\tilde{\mathbf{r}}(u)\right)\right| du. \tag{5.1}$$

The tangent vector $\mathbf{t}(s)$, a unit vector that is tangent to this curve, has the form

$$\mathbf{t}(s) \equiv \frac{1}{\sqrt{(d\tilde{x}/du)^2 + (d\tilde{y}/du)^2}}\left(\frac{d\tilde{x}}{du}, \frac{d\tilde{y}}{du}\right)$$

$$= \frac{d}{ds}\mathbf{r}(s), \tag{5.2}$$

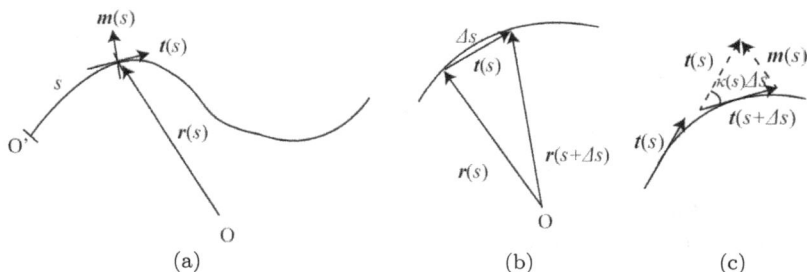

Fig. 5.1: A 1D curve embedded in 2D flat surface is represented as a positional vector that is a function of contour length s (a). The tangent vector $\mathbf{t}(s)$ is a unit vector that is tangent to this curve and the normal vector $\mathbf{m}(s)$ is a unit vector that is normal to the tangent curve. The tangent curve is a unit vector because $|\mathbf{r}(s + \Delta s) - \mathbf{r}(s)|/\Delta s \to 1$ for $\Delta s \to 0$ (b). Tangent vectors are unit vectors, and the derivatives of tangent vectors are normal to the tangent vector (c). $\kappa(s)\Delta s$ is the angle made by the tangent vector at s and the tangent vector at $s + \Delta s$; curvatures are extent that tangent vectors rotate with increasing s.

where $r(s)$ is the positional vector represented as a function of s; s is the contour length and used as the parameter of curves. Taking derivatives to $r(s)$ leads to unit vector $t(s)$ that are tangent to the curves, where $\lim_{\Delta s \to 0} \frac{|\Delta r(s)|}{\Delta s} = 1$ (see Fig. 5.1(b)).

The tangent vectors are unit vectors, $t(s) \cdot t(s) = 1$, and taking derivative with respect to s leads to $dt(s)/ds \cdot t(s) = 0$; the derivatives of tangent vectors $dt(s)/ds$ are normal to the tangent vectors $t(s)$. The derivatives of tangent vectors thus have the form;

$$\frac{d}{ds} t(s) = \kappa(s) m(s), \tag{5.3}$$

where $\kappa(s)$ is the curvatures of curves (see also Fig. 5.1(c)). $m(s)$ is the normal vector, a unit vector normal to the tangent vector $t(s)$, and is the outward of domains [2, 3];

$$m(s) = t(s) \times n, \tag{5.4}$$

where n is normal to the 2D flat surface (n represents the interface normal and thus this notation is consistent with domains in monolayers, where the normal direction is perpendicular to the flat surface). Taking the derivative of Eq. (5.4) with respect s using Eq. (5.3) leads to the form

$$\frac{d}{ds} m(s) = -\kappa(s) t(s). \tag{5.5}$$

Equations (5.3) and (5.5) are summarized in the form of a first-order ordinary differential equation

$$\frac{d}{ds} \begin{pmatrix} t(s) \\ m(s) \end{pmatrix} = \begin{pmatrix} 0 & \kappa(s) \\ -\kappa(s) & 0 \end{pmatrix} \begin{pmatrix} t(s) \\ m(s) \end{pmatrix}. \tag{5.6}$$

When the forms of curvatures $\kappa(s)$ as functions of contour length are given, in principle, Eq. (5.6) can be solved with a boundary condition that specifies the orientation of the curve. The solution $t(s)$ is further integrated by using Eq. (5.2) with another boundary condition that specifies the position of the curve; 1D curves embedded in 2D flat surfaces are determined by their curvatures $\kappa(s)$, except for their orientations and translations.

It may be worthwhile to note that κ is included in the asymmetric tensor on the right-hand side of Eq. (5.6). This implies that tangent and normal vectors, $\mathbf{t}(s)$ and $\mathbf{m}(s)$, rotate by angle $\kappa(s)\Delta s$ as one moves along the 1D curve by Δs; $\mathbf{t}(s)$ and $\mathbf{m}(s)$ define a local coordinate system along this curve (see also Fig. 5.1(c)). This formalism is called Frenet–Serret frame or, simply, Frenet–Serret theorem [2].

Noteworthy that, according to the definition of vectors, $\mathbf{t}(s)$ and $\mathbf{m}(s)$, these unit vectors can be represented in another way as follows:

$$t(s) = (\cos \phi(s), \sin \phi(s)),$$

and

$$m(s) = (\sin \phi(s), -\cos \phi(s)), \tag{5.7}$$

where $\phi(s)$ is the boundary orientational angle at s and is related to the curvature by

$$\kappa(s) = -\frac{d\phi(s)}{ds}. \tag{5.8}$$

5.2. Modeling of Domain Shapes in Monolayers on Water Surface

5.2.1. *Domain shapes and modeling*

Monolayers of lipid on the water surface have been studied for more than a century since the discovery of Langmuir monolayers [4]. The aspect of different phases and their coexistence in these two-dimensional (2D) systems has been expected in analogy with 3D ones [5,6]. Such predictions have been directly visualized by using fluorescence microscopy (FM) [7–9], Brewster Angle microscopy (BAM) [10,11], etc. Interestingly, monolayers stabilize domains of condensed phase that coexist with domains of liquid expanded phase, and these condensed phase domains show intriguing shapes that are very different from 3D materials. The microscopic observation has revealed a variety of ordered and disordered domain shapes, and interesting behaviors of monolayers [12]. For example, domains in dipalmitoyl-phosphatidylcholine (DPPC) monolayers show circular and clover shapes (see Fig. 5.2 [12–16]). Of course, not only circular and clover leaf shapes, but also m-sided domains ($m = 1, 2, 3, 4, 5\ldots$), etc.

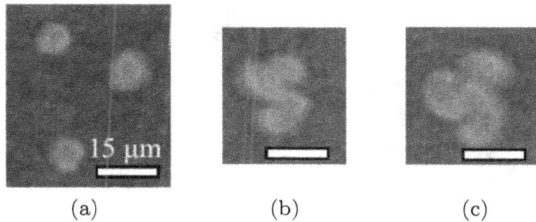

(a)   (b)   (c)

Fig. 5.2: Shapes of domains of condensed phase that coexist with domains of liquid expanded phase in D-dipalmitoyl-phosphatidylcholine (DPPC) monolayers. These domains show circular domains at relatively small surface pressures Π (a) and show shape transitions to clover-shaped domains, (b) and (c). The number of lobes increases with increasing surface pressures Π. Micrograph (a) is captured at $\Pi = 7.93 \, \text{mN/m}$ and micrographs (b) and (c) are captured at $\Pi = 10.3 \, \text{mN/m}$. [Reproduced from Ref. [14].]

are observed, depending on the surface pressure applied to mono-layers, etc. (see Fig. 5.2 [14, 17]). It is instructive here to note that DPPC molecules are chiral molecules and thus have two chi-ral isomers (so-called enantiomer), where the molecular structure of one isomer is the mirror image of the other. The lobes of clover-shaped domains in DPPC monolayers are bent toward clockwise or anti-clockwise directions depending on the chirality of constituent molecules (see Fig. 1.4); domains of one enantiomer show shapes that are mirror images of domains of the other. We here call shapes that depend on the chirality of constituent molecules as *chiral shapes*. Indeed, racemic monolayers that are one-to-one mixtures of two enantiomers show clover-shaped domains that have mirror symmet-ric shapes (Fig. 5.3) [13, 15, 16]. These visualized images show that the shapes of domains in monolayers are sensitive to the chirality of constituent molecules. It is of interest to elucidate the physical mech-anisms involved in stabilizing these characteristic shapes of domains in monolayers.

Among a variety of domain shapes, circular shapes of solid domain surrounded by fluid phase, and deformed shapes of their transi-tions to torus [18], D form, S form, serpentine manner form [19, 20], and m-sided quasi-polygon form [21, 22] will be helpful to study the physics on 2D domain shapes [22–25]. It is instructive here to note that there are at least two independent approaches for study-ing domain shapes. The first one is to understand domain shapes

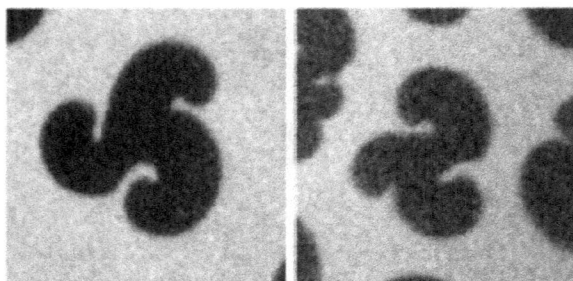

Fig. 5.3(a): Domains of L-DPPC and D-DPPC on pure water (T ≈ 20°C, $\pi \approx 9\,\text{mN/m}$) as visualized with Fluorescence micrograph (FM) ~0.5-mol% NBD-PC).
Source: Reprinted from Figs. 1 and 3 in Ref. [15], *Physical Rev. E*, 62, 7031 (2000).

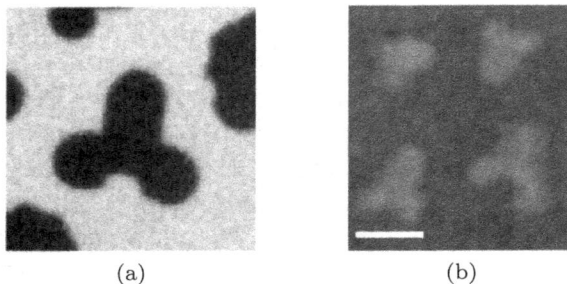

(a)   (b)

Fig. 5.3(b): Domain shapes of racemic DPPC monolayer on water surface at 21°C, collected by BAM at (a) $2.27\,\text{mN/m}$ (68.5 Å^2) and (b) $2.66\,\text{mN/m}$ (62.0 Å^2), during the compression rate (0.74 Å^2/min a molecule).
Source: Reprinted from Ref. [16], *Colloids Surf. A Physicochem. Eng. Aspects*, 321 151–157, 2008.

from the viewpoint of the peculiar polar property of the monolayers of amphiphilic molecules at the air–water interface; monolayers comprise non-symmetric dipolar molecules and solid domains is formed due to the competition between line tension and electrostatic repulsion. This is the basic approach proposed by McConnell [24]. The second one is to understand domain shapes by focusing on the inner textures of domains, i.e., their crystalline nature (see Chapters 6 and 7); this leads to the idea that starts from the two-dimensional (2D) Frank elasticity theory of liquid crystals. In this chapter, we mainly focus on the former approach, where the constituent molecules of monolayers are simply modeled as rod-like dipolar molecules. In the following chapters, we also discuss the analyses with consideration of the latter idea.

Experimentally, we know that monolayers are composed of amphiphile molecules, and exhibit *in-plane* orientational order in condensed phase possibly due to the liquid-crystalline nature of monolayers. Consequently, monolayer domains have the possibility to possess *in-plane* spontaneous polarizations. The mathematics to treat 2D domains with *in-plane* electric dipoles is rather complex. As that, we first focus on domains with no *in-plane* electric dipoles, but this does not lose the generality of physics on 2D domain shapes. Indeed, we can see such domains on water surface. After that we treat the case with *in-plane* electric dipoles, by extending the mathematics used for analyzing domains without *in-plane* electric dipoles. The fact that monolayers show non-centrosymmetric orientational structures with electric dipoles normal to the water surface, but without *in-plane* electric dipoles, implies that the electric dipoles of constituent molecules show long-range repulsive electrostatic dipole–dipole interactions; these interactions obviously store electrostatic energy in the domains of monolayers. Electrostatic energy arising from the interactions between a pair of electric dipoles has the form of Eq. (2.12). Electrostatic energy stored in a domain in monolayers is just the sum of Eq. (2.12) for all the pairs of electric dipoles in the domain. In the spirit of continuum theories of dielectric polarizations, we treat normal electric dipoles that are averaged over the area that is larger than the molecular length scale and is smaller than the size of domains, see the treatments in Sections 2.1.2 in Chapter 2 and Appendix B. In this description, electrostatic energy arising from dipole–dipole interactions has the form

$$F_{\perp} = \frac{1}{4\pi\epsilon_0} \frac{u_{\perp}^2}{2} \int dS_i \int dS_j \frac{1}{|\boldsymbol{R_i} - \boldsymbol{R_j}|^3}, \qquad (5.9)$$

where $\mu_{\perp}$ is the area density of electric dipoles, normal to the water surface. $\boldsymbol{R_i}$ and $\boldsymbol{R_j}$ are positional vectors in the domain, and dS_i and dS_j are the area elements at these positions (the subscripts i and j represent the elements in domains because two area integrals over two positional vectors are involved in the calculations of Eq. (5.9)).

5.2.2. *McConnell' approach*

McConnell and coworkers proposed that the shapes of domains in monolayers are determined as the minimum state of the free energy given by [24]

$$F = \lambda_0 \oint ds + F_\perp, \tag{5.10}$$

where λ_0 is line tension, and ds is line element along the boundary of the domain. Line tension λ_0 is the free energy cost to make an interface of unit length and the statistical thermodynamics of interfacial tensions is shown in Safran's book [25]. Intuitively, line tension tends to make the domain shape circular to minimize the length of the boundary. On the other hand, electrostatic dipolar energy tends to elongate the domain shapes because long-range repulsive dipole–dipole interactions tend to separate two arbitrary pieces of the domain as far as possible. Consequently, the shapes of domains are modeled to be determined by the competition between them. The physical concept of this McConnell's idea is relatively simple, but to treat Eq. (5.10) we need to carry the double integral of electrostatic dipolar energies given by Eq. (5.9).

5.2.3. *Derivation of free energy F*

Equation (5.9) is further simplified by using a mathematical formula

$$\frac{1}{|\mathbf{R}_i - \mathbf{R}_j|^3} = \frac{2\pi}{\Delta} \delta(|\mathbf{R}_i - \mathbf{R}_j|) + \boldsymbol{\nabla}_{\|j} \cdot \left(\frac{\mathbf{R}_i - \mathbf{R}_j}{|\mathbf{R}_i - \mathbf{R}_j|^3} \right), \tag{5.11}$$

where $\boldsymbol{\nabla}_{\|j}$ is 2D gradient along the monolayer surface with respect to the 2D positional vector $\mathbf{R}_j$. Equation (5.11) is only valid for 2D surfaces. Δ is the cutoff distance because of the fact that the second term of Eq. (5.11) is not differentiable at $\mathbf{R}_i = \mathbf{R}_j$. We here remind that electrostatic energy arising from dipole–dipole interactions has the form of Eq. (5.9), only for distances $|\mathbf{R}_i - \mathbf{R}_j|$ that are larger than the half of the length of electric dipoles. We thus use the half of monolayer thickness $h/2$ as cutoff Δ. Using Eq. (5.11) with Storks' theorem

$$\int dS \, \mathbf{n} \cdot \boldsymbol{\nabla}_\| \times \mathbf{A}(\mathbf{R}) = \oint ds \, \mathbf{t}(s) \cdot \mathbf{A}(\mathbf{r}(s)), \tag{5.12}$$

Equation (5.9) is rewritten in the form

$$F_\perp = \frac{\mu_\perp^2}{2} \frac{2\pi}{\Delta} \int dS - \frac{\mu_\perp^2}{2} \oint ds_i \oint ds_j \frac{\mathbf{t}(s_i) \cdot \mathbf{t}(s_j)}{|\mathbf{r}(s_i) - \mathbf{r}(s_j)|}, \qquad (5.13)$$

where $\mathbf{r}(s)$ is the positional vector at the boundary of the domain and s is contour length. The first term of Eq. (5.13) is electrostatic energy generated when charges $\frac{\mu_\perp}{h} \int dS$ are stored in a parallel capacitor with capacitance $\frac{1}{4\pi h} \int dS$, where h is the distance between two parallel electrodes; this implies that the cutoff Δ is $h/2$. Indeed, the second term of Eq. (5.13) is the electrostatic energy arising from fringe electric fields that emanate from the capacitance at its boundary. McConnell and coworkers neglected the first term of Eq. (5.13) and rewrote the free energy given by Eq. (5.10) in the form

$$F = \lambda_0 \oint ds - \frac{\mu_\perp^2}{2} \oint ds_i \oint ds_j \frac{\mathbf{t}(s_i) \cdot \mathbf{t}(s_j)}{|\mathbf{r}(s_i) - \mathbf{r}(s_j)|}. \qquad (5.14)$$

5.2.4. *Analysis under assumption of simple domain shape form*

One of the conventional ways to find solutions that satisfy minimum energy of Eq. (5.14) is to assume most probable solutions *a-priori*, and check the validity of the assumed solutions. McConnell and coworkers analyzed the shapes of domains by assuming that the domain has the symmetric shape with the form (see, e.g., Fig. 5.2(a))

$$\mathbf{r}(s) = r_0(\cos\theta, \sin\theta) + g(\theta)(\cos\theta, \sin\theta), \qquad (5.15)$$

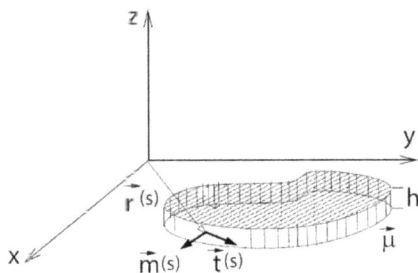

Fig. 5.4:   Schematic illustration of the shape of a 2D monolayer domain, whose dipoles μ are normal to the x–y plane ($\mu = \mu_\perp$). The geometric quantities describing its boundary curve are also shown, and h is the thickness of the monolayer.
Source: Reprinted from Fig. 1 in Ref. [39], M. Iwamoto, F. Liu, and Z. C. Ou-Yang, *J. Chem. Phys.*, 125, 224701 (2006).

where θ is the polar angle from the x-direction (that is defined in an arbitrary direction). For the case of $g(\theta) = 0$, Eq. (5.15) represents circular domains, i.e., the most probable domain shape, and its free energy given by Eq. (5.14) has the form

$$F = 2\pi r_0 \left[\lambda_0 + \mu_\perp^2 \ln \left(\frac{e^2 \Delta}{8 r_0} \right) \right], \tag{5.16}$$

where e is the base of natural logarithm.

Minimizing Eq. (5.16) with respect to r_0 leads to a stable radius of circular domain in the form of $r_{eq} = \frac{e\Delta}{8} e^{\lambda_0/\mu_\perp^2}$. That is, the existence of circular domains in equilibrium is predicted in terms of a competition between line tension and long range electro-static force induced by molecular dipoles in domains, i.e., minima of the domain energy.

The stability of the circular domains can be analyzed by calculating the free energy by the second-order terms with respect to $g(\theta)/r_0$, meanwhile the area of the domain is fixed to the stable area, πr_{eq}^2.

Because the boundary line of the domain is a closed curve, $g(\theta)$ is expanded in a Fourier series

$$g(\theta) = \sum_{n=-\infty}^{\infty} c_n e^{in\theta}, \tag{5.17}$$

with $c_{-n} = c_n^*$, where the asterisk "$*$" represents the complex conjugate. A straight forward calculation leads to the form of the free energy of domains as

$$\delta F = \frac{\pi \mu_\perp^2}{r_0} \sum_{n=-\infty}^{\infty} \Omega_n^\perp |c_n|^2, \tag{5.18}$$

with

$$\Omega_n^\perp = \frac{\lambda_0}{\mu_\perp^2} (n^2 - 1) - \frac{(2n+1)^2}{16} L_{n+1} - \frac{(2n-1)^2}{16} L_{n-1}$$

$$+ \frac{3}{8} L_1 + \frac{1}{8} \left(M_1 - \frac{M_{n+1} + M_{n-1}}{2} \right), \tag{5.19}$$

by the second-order term with respect to c_n/r_0. Here, L_n and M_n are integrals of the form

$$L_n = \int_{-\pi}^{\pi} d\theta \frac{\cos n\theta}{\sqrt{1 - \cos^2 \frac{\theta}{2} + \frac{\Delta^2}{4 r_0^2}}} \quad \text{and} \quad M_n = \int_{-\pi}^{\pi} d\theta \frac{\cos n\theta}{\left(1 - \cos^2 \frac{\theta}{2} + \frac{\Delta^2}{4 r_0^2} \right)^{3/2}}.$$

In the derivation of Eq. (5.18), the following condition of constant domain area is used (see Section 5.4.3.2).

$$\frac{c_0}{r_0} = \frac{1}{2} \sum_{n=-\infty}^{\infty} \frac{|c_n|^2}{r_0^2}. \tag{5.20}$$

Note that Deutch and Low started from Eq. (5.10) and obtained the same result [26].

Equation (5.19) includes integrals L_n and M_n that are not evaluated. Miranda used the integral form of the Legendre polynomials of second kind and derived an expression of $\Omega_n^\perp$ in the form [27]

$$\Omega_n^\perp = \frac{\lambda_0}{\mu_\perp^2}(n^2 - 1) + (n^2 - 1)\log\left(\frac{e^2\Delta}{8r_0}\right) + \frac{n^2 - 1}{2}$$
$$- \frac{4n^2 - 1}{4}\left(\psi\left(\frac{3}{2}\right) - \psi\left(n + \frac{1}{2}\right)\right), \tag{5.21}$$

where $\psi(n)$ is digamma function. Equation (5.21) is evaluated by using the relationship

$$\psi\left(n + \frac{1}{2}\right) - \psi\left(n - \frac{1}{2}\right) = \frac{2}{2n - 1}. \tag{5.22}$$

Equation (5.18) implies that circular domains are stable when the coefficients $\Omega_n^\perp$ are positive for all values of n, whereas the circular domains are no longer stable when one of the coefficients $\Omega_n^\perp$ is negative. For a domain radius r_0 that is smaller than a threshold value, $r_2 = \frac{\Delta}{e}e^{\lambda_0/\mu_\perp^2 - 10/3}$, circular domains are stable. Circular domains are no longer stable for a domain radius that is larger than r_2 and show "shape transitions" to clover-shaped domains with n-lobes when $\Omega_n^\perp$ is negative, see also Eq. (5.15). Because $\Omega_n^\perp$ is an increasing function of n, see Eq. (5.21), domains stabilize from clover-shaped domains of fewer lobes as one increases the radius of the domain. $\Omega_n^\perp$ is zero for $n = 1$ because this mode, c_1, corresponds to translations of domains. Threshold radii that are predicted by this theory semi-quantitatively agree with the results of experiments on domains in monolayers that show liquid–liquid phase separations. This theory predicts that monolayers stabilize clover-shaped domains and thus, in principal part agrees with the experimentally observed shapes of domains in DPPC monolayers. However, the predicted clover-shaped domains have straight lobes, and this is contradictory to visualized clover-shaped domains in DPPC monolayers that have bent lobes (see Figs. 5.2 and 5.3).

Because the directions, in which these lobes of clover-shaped domains are bent, reflect the effect of chirality on the shapes of domains in monolayers, the physical mechanisms involved in the bending of these lobes of clover-shaped domains are still not clearly understood only by the McConnell's approach. Nevertheless, the fact that the electrostatic analysis by using the McConnell's approach can predict the stability of circular domains and the growth of clover-shaped domains implies that electrostatic energy arising from dipole–dipole interactions play an important role in the formation and evolution of intriguing shapes of domains in monolayers. And the shapes of domains in equilibrium and their deformation could not be understood without consideration of a competition between line tension and long range electro-static force induced by molecular dipoles in domains, i.e., minima of the domain energy.

As mentioned above, the treatment of domain shapes of monolayers by McConnell's approach is instructive, though not complete. The main reason of the incompleteness is that this approach starts under assumption that domain shapes are given by Eq. (5.15), though a variety of domain shapes which are totally different from circular and related shapes, have been visualized experimentally. Nevertheless, the conceptional idea that the shapes of domains are governed by the competition between line tension and long range electro-static force induced by molecular dipoles in domains, i.e., minima of the domain energy, is very instructive to study domain shapes.

5.3. Free Energy of Domains in Monolayers

5.3.1. *History of the geometric study on shapes of materials*

As mentioned in Section 5.2, the pioneering work by the McConnell group is instructive. However, for further understanding stabilized domain shapes on the water surface, the mathematical knowledge on

Differential geometry of 2D flat surfaces (see Section 5.1) is helpful and necessary. The derivation of shape equations that are described by using geometric parameters, without assuming domain shapes *a priori*, is effective.

Before going to the derivation of such shape equations, it is instructive here to review the history of the geometric study on shapes of materials. Briefly, shape problems stem from real interfaces of crystals, liquids, and biological systems in nature. Since the pioneering studies of soap bubbles by Plateau [28] and the shape of interfaces in a capillary tube by Young [29] and Laplace [30] in the 19th century, minimal surfaces or the variation problem $\delta \oint dA = 0$ and surfaces with constant curvature, $\delta \left[\Delta P \int dV + \lambda \oint dA \right] = 0$, have initiated a first "golden age" in differential geometry, where ΔP is the pressure between air/liquid interface, λ is the surface tension, A is the area of the interface, and V is the volume of liquid phases. By generalizing the surface tension to anisotropic case, i.e., λ depending on the direction n of crystal surface, the variation equation of $\delta \left[\Delta P \int dV + \oint \lambda(\boldsymbol{n}) dA \right] = 0$ yields the famous Wolf construction [31], which successfully predicts convex and faceted shapes of equilibrium crystals. Nowadays, such kind of variations have been developed to include the integral of the mean curvature H and Gaussian curvature K of the interface, for instance, the important variation equation of the Helfrich free energy F_H [32].

$$F_H = \Delta P \int dV + \lambda \oint dA + \overline{k_c} \oint K dA + (k_c/2) \oint (2H + C_0)^2,$$

$$(5.23)$$

where k_c and $\overline{k_c}$ are bending rigidity and Gaussian curvature modulus, respectively, and C_0 the spontaneous curvature. That is, the Helfrich free energy is described by using geometrical parameters, such as the mean curvature H and Gaussian curvature K. After the variation, the equation $\delta F_H = 0$ that satisfies minimum Helfrich free energy condition, i.e., the three-dimensional (3D) shape equation, has been derived as follows [33]:

$$\Delta P - 2\lambda H + 4k_c \left(H + \frac{C_0}{2} \right) \left(H^2 - K - \frac{C_0}{2} H \right) + 2k_c \nabla^2 H = 0.$$

$$(5.24)$$

The solution of this shape equation can analytically predict the well-known shapes of vesicles formed by lipid bilayer in aqueous, e.g., the biconcave–discoid shape of red blood cells [34] and the Clifford torus, namely a torus with two generating circles of radii r and R with $r/R = 1/\sqrt{2}$ [35–37], and so on. Therefore, we anticipate that the extension of 3D Helfrich variation problem is one of most promising key steps to resolve 2D domain problems, e.g., shapes of domains in monolayers.

5.3.2. *Reconsideration of expression of free energy formula of domains*

Shape equation of the domains will be determined by the variation of the McConnell free energy of Eq. (5.14). However, at first glance, Eq. (5.14) is distinctly different from the Helfrich free energy form given by Eq. (5.23) in that Eq. (5.14) comprises line tension energy and electric dipolar interaction energy. As that, we need to reconsider the form of free energy of two-dimensional domains. Keeping in mind the form of Helfrich curvature-elastic energy, a Lagrange multiplier ΔP is additionally introduced, with consideration of the constraint of constant domain area and/or the difference in the Gibbs free energy density between outer (e.g., fluid) and inner (solid) phases, i.e., $\Delta P = -g_0$. Here $g_0 > 0$, since solid phase is more stable than fluid one. Actually, advanced microscope observation shows that solid domains (LC phase) are surrounded by fluid phase (LE phase) in monolayers (see Chapter 1).

Then the shape energy of a 2D domain of Eq. (5.14) is replaced as [38, 39]

$$F = \Delta P \int dA + \lambda \oint ds - \frac{\mu_\perp^2}{2} \oint \oint \frac{\mathbf{t}(l) \cdot \mathbf{t}(s)}{|\mathbf{r}(l) - \mathbf{r}(s)|} dl ds, \qquad (5.25)$$

where λ is the line tension, $\mu_\perp$ is the dipole density, $\mathbf{r}(s)$ describes the domain boundary curve; s is the arclength; and $\mathbf{t} = d\mathbf{r}/ds$ is the unit tangential vector of the boundary (see Fig. 5.3).

This equation looks similar to the equation of Helfrich free energy. But instead of the curvature-elastic energy F_c [sum of the 3rd and 4th terms of Eq. (5.23)], the origin is of course the electrostatic dipole–dipole interaction energy [the last term of (5.25)]. Therefore, we must show that the electrostatic dipole–dipole interaction energy term can be converted to the form corresponding to the curvature-elastic energy. The long-lasting insufficient stage of the 2D domain theory might be due to the mathematical difficulty in calculating the double line integral, the last term of Eq. (5.25), and rewriting Eq. (5.25) in an analogous form to the curvature elastic energy form by Helfrich. To do this, employing Taylor approximation to the dipolar energy term will be helpful. In Section 5.4, we show that the double line integral of the dipolar energy can be approximately derived as a sum of an additionally negative line tension and a curvature-elastic energy of the domain boundary, by employing the Taylor approximation method. As the result, derived expression becomes very similar to the form of curvature-elastic energy by Helfrich, and enables us to calculate the δF analytically, to derive equations expressed using geometric parameters.

Using the new expression form of Free energy, a variety of domain shapes can be predicted as its solutions, depending on ΔP, total line tension, and the dipole force-induced curvature elastic-modulus. Further, it can be shown that the infinitesimal instability of a circle with fixed area deforms the circle into a shape associated with m-th order harmonics at below some negative threshold line tension. Of course, the comparison of the shapes with the experimental observation domains will be of importance to confirm the validity of the method using *differential geometry*.

5.4. Derivation of Approximate Shape Equation of Domains in Monolayers without *in-plane* Electric Dipoles [38, 39]

5.4.1. *Free energy of domains in monolayers without in-plane electric dipoles*

In the derivation, the first key step is to rewrite the dipolar force energy as

$$F_\perp = -\frac{\mu_\perp^2}{2} \oint \left(\oint \frac{\mathbf{t}(s) \cdot \mathbf{t}(s+x)}{|\mathbf{r}(s+x) - \mathbf{r}(s)|} dx \right) ds, \qquad (5.26)$$

where arc-variable x is defined as $x = l - s$. By employing Frenet formulae of a plane curve [30] (see Section 5.1), one has

$$\mathbf{r}(s+x) = \mathbf{r}(s) + \mathbf{t}(s)x + \frac{1}{2}\kappa(s)\mathbf{m}(s)x^2$$

$$+ \frac{1}{6}(\kappa_s(s)\mathbf{m}(s) - (s)\mathbf{t}(s))x^3 + \qquad (5.27)$$

$$\mathbf{t}(s+x) = \mathbf{t}(s) + \kappa(s)\mathbf{m}(s)x$$

$$+ \frac{1}{2}(\kappa_s(s)\mathbf{m}(s) - \kappa^2(s)\mathbf{t}(s))x^2 + \cdots, \qquad (5.28)$$

where $\kappa(s)$ and $\mathbf{m}(s)$ are the curvature and unit normal vector of the boundary curve at s. Substituting (5.27) and (5.28) into (5.26) gives the expression of the dipole force energy in a generalized line tension energy form as $F_\perp = \oint \overline{\lambda}(\kappa, \kappa_s, \ldots)ds$, where $\overline{\lambda}$ is a series of $\kappa(s)$ and its derivatives. If we consider a somewhat smooth curve, and only take account of the first two terms of $F_\perp$ series as Taylor approximation and using the monolayer thickness h, nonzero, as the cutoff Δ [40], we can approximately obtain the expression given by

$$F_\perp = -\frac{\mu_\perp^2}{2} \oint ds \int_0^L dx \frac{\mathbf{t}(s) \cdot \mathbf{t}(s+x)}{|\mathbf{r}(s+x) - \mathbf{r}(s)|}$$

$$\simeq -\frac{\mu_\perp^2}{2} \oint ds \int_h^L dx \left[\frac{1}{x} - \frac{11}{24}\kappa(s)^2 x + o(x^2)\right]$$

$$= -\frac{\mu_\perp^2}{2} \log\frac{L}{\Delta} \oint ds + \frac{11}{96}\mu_\perp^2 L^2 \oint ds\kappa^2(s). \qquad (5.29)$$

Here we omitted the higher order terms with respect to curvature $\kappa(s)$. Because if the free energy, Eq. (5.23), is rewritten in the form of curvature elastic energy, it is enough to cut a series of Taylor expansion terms by the terms that are proportional to $\kappa^2(s)$. In principle, expansions by using Eq. (5.27) and (5.28) are good approximations only for x that is smaller than κ^{-1}; and using Eqs. (5.27) and (5.28) even for $x \sim L$ is an approximate treatment.

Equation (5.29) shows that electrostatic energy arising from dipole–dipole interactions has an approximate form of the sum of negative line tension and curvature elastic energy. With this approximate form, the free energy of domains in monolayers is described as

$$F = \Delta P \int dS + \tilde{\lambda} \oint ds + \alpha_{\perp} \oint ds \kappa^2(s), \qquad (5.30)$$

with $\tilde{\lambda} = \lambda - \frac{\mu_{\perp}^2}{2} \ln(L/\Delta)$ and $\alpha_{\perp} = \frac{11}{96}\mu_{\perp}^2 L^2$. Here $\Delta P (\equiv g_{in} - g_{ex})$ is the difference of Gibbs free energy (area) density between domain interior and exterior (or the difference of surface pressures, $\Pi_{ex} - \Pi_{in}$, between domain interior and exterior). It is of interest to note that the form of Eq. (5.30) is analogous to the curvature elastic energy F_H of 3D lipid vesicles by Helfrich, i.e., Eq. (5.23). In other words, the double line integral of the dipolar energy can be approximately expressed as a sum of an additionally negative line tension and a curvature-elastic energy of the domain boundary.

Noteworthy that a clear physical basis has not yet been provided for the Taylor expansion of the dipolar interaction; however, it is efficient in mathematics when the ratio of the distance between two adjacent dipoles to the radius of domain boundary is small. As observed in experiment, the radius of domain is on the order of micro-meters and quite larger than the distance of adjacent dipoles. Thus, it is always satisfied that the parameter in expansion is small. The validity of such approximation is as follows: If we consider the limit $h \to 0$ the divergence of Eq. (5.29) depends only on the integral of the first integrand term of $1/x$; it is a good approximation to neglect the terms of $O(x^2)$. For an exact calculation for a circle of radius ρ_0, the logarithmic divergence is calculated as $\ln\left(8\rho_0/e^{\frac{1}{2}}h\right) \approx \ln(5\rho_0/h)$ (see (2.11) in Ref. [29]), and this is nearly equal to $\ln\frac{2\pi\rho_0}{h}$ of the current calculation result with the Taylor expansion approach. In fact, such an *ansatz* approach by expanding the line integral was also proposed by Langer, *et al.* (see, (5 11) and (5.12) in Ref. [40]) without further justification.

Although the McConnell group has provided sound theoretical calculation to explain the domain shape bifurcation diagram for somewhat regular shapes [23, 24] (see Section 5.2), the expression (5.30) can extend the study to general shapes. Interestingly, one can find from (5.30) that 2D line tension is size- and shape-dependent form. This is the most important characteristics of the free energy of 2D domains in monolayer, leading to a variety of domain shapes. In other words, electrostatic dipolar energy is converted to a negative line tension and a curvature-elastic energy, accordingly, and leading to a variety of intriguing domain shapes.

5.4.2. Derivation of shape equation of domains with no in-plane electric dipoles

The variational principle is used to derive a shape equation that has the stable shapes of domains in monolayers in the solution. With a small variation $\Gamma(s)$, the shape of a domain changes to a new shape that is described by the positional vector

$$\mathbf{r}'(s') = \mathbf{r}(s) + \Gamma(s)\mathbf{m}(s). \tag{5.31}$$

Here the symbol prime "$'$" stands for the geometrical quantities of the new shape. The s' is the contour length of the new shape, and the s no longer represents the contour length of the new shape. We thus use Eq. (5.1) to derive ds' in the form

$$ds' = \sqrt{(1 - \kappa(s)\Gamma(s))^2 + \Gamma_s^2}\,ds \simeq (1 - \kappa(s)\Gamma(s))ds, \tag{5.32}$$

where Γ_s is the first derivative of $\Gamma(s)$ with respect to s. The tangent and normal vectors of the new shape are derived by using Eqs. (5.2) and (5.4), and have the forms

$$\mathbf{t}'(s') = \mathbf{t}(s) + \Gamma_s\mathbf{m}(s), \tag{5.33}$$

$$\mathbf{m}'(s') = \mathbf{m}(s) - \Gamma_s\mathbf{t}, \tag{5.34}$$

by the first order term with respect to $\Gamma(s)$. The curvatures of this new shape have the form

$$\kappa'(s') = \kappa + \kappa^2\Gamma + \Gamma_{ss}, \tag{5.35}$$

where Γ_{ss} is the second derivative of $\Gamma(s)$ with respect to s.

The free energy, Eq. (5.30), of domains is a functional of $\Gamma(s)$. The variation of the first term of Eq. (5.30) is calculated in the form

$$\delta^{(1)}[\lambda \oint ds] \equiv \lambda \oint ds' - \lambda \oint ds$$

$$= \oint ds(-\lambda\kappa)\Gamma(s), \tag{5.36a}$$

by the first-order term with respect to $\Gamma(s)$. $\delta^{(1)}[F[\mathbf{r}(s)]](\equiv F[\mathbf{r}'(s')] - F[\mathbf{r}(s)])$ is the first order variation of a functional $F[\mathbf{r}(s)]$ with respect to $\Gamma(s)$. Expanding the other terms of Eq. (5.30) leads to

$$\delta^{(1)}\left[-\frac{\mu_\perp^2}{2}\oint ds\ln\frac{L}{\Delta}\right] = -\frac{\mu_\perp^2}{2}\ln\frac{eL}{\Delta}\oint ds(-\kappa)\Gamma(s), \qquad (5.36\text{b})$$

$$\delta^{(1)}\left[\frac{11}{96}\mu_\perp^2 L^2\oint ds\kappa^2\right] = \left(\frac{11}{48}\mu_\perp^2 L^2\oint ds\kappa^2\right)\oint ds\kappa^2$$

$$+\frac{11}{96}\mu_\perp^2 L^2\oint ds\Gamma(s)(\kappa^2+2\kappa_{ss}). \qquad (5.36\text{c})$$

We here note that $\tilde{\lambda}$ and $\alpha_\perp$ are functions of the boundary length L, and the variation of these coefficients must be considered. The first order variation of the free energy, Eq. (5.30), thus has the form

$$\delta^{(1)}F = \oint ds\Gamma[\Delta P - \lambda_{\text{ele}}\kappa + \alpha_\perp\kappa^3 + 2\alpha_\perp\kappa_{ss}], \qquad (5.37)$$

with

$$\lambda_{\text{ele}} = \lambda - \frac{\mu_\perp^2}{2}\ln\frac{eL}{\Delta} + \frac{11}{48}\mu_\perp^2 L^2\oint ds\kappa^2.$$

Here, κ_{ss} is the second derivative of κ with respect to s.

For the shapes of minimal free energy, the first order variation of the free energy functional is zero (for arbitrary functions of $\Gamma(s)$), $\delta^{(1)}F = 0$; this is a similar logic to the fact that the first derivatives of functions are zero at their minima. We thus get the shape equation in an approximate form as

$$\Delta P - \lambda_{\text{ele}}\kappa + \alpha_\perp\kappa^3 + 2\alpha_\perp\kappa_{ss} = 0. \qquad (5.38)$$

By solving this derived equation, shapes of domains can be predicted.

5.4.3.   *Solution of shape equation: Circular domains*

5.4.3.1.   *Circular domains*

Equation (5.38) is the shape equation expressed using geometric quantities κ and κ_{ss}. By solving this equation, we can discuss shapes of stable domains that will be visualized on water surfaces. It is of interest to note that Eq. (5.38) has the analogous form to the shape equation of lipid vesicles (see Eq. (5.24)). Note that this equation is derived by defining the normal direction to be outward, $\kappa = -1/\rho_c$ for a circle with radius ρ_c. According to this definition, $\kappa = 1/\rho_t$ is then the curvature of the inner circle of a torus with radius ρ_t. Given these, it is possible to represent shapes with coexisting outer and inner circles if $\rho_t < \rho_c$.

Keeping into mind these relations and the experimental evidence of the presence of circular domains, it is of interest to analyze the case of $\kappa_{ss} = 0$, using Eq. (5.38). By rewriting Eq. (5.38) as a function $\Delta P(\kappa) = \lambda \kappa - \alpha \kappa^3$ (here we represent $\lambda_{\text{ele}} = \lambda$ and $\alpha_\perp = \alpha$ for simplicity) considering $\alpha > 0$, we illustrate the behavior of ΔP in Fig. 5.5(a) and (b) for $\lambda > 0$ and $\lambda \le 0$, respectively. From Fig. 5.5,

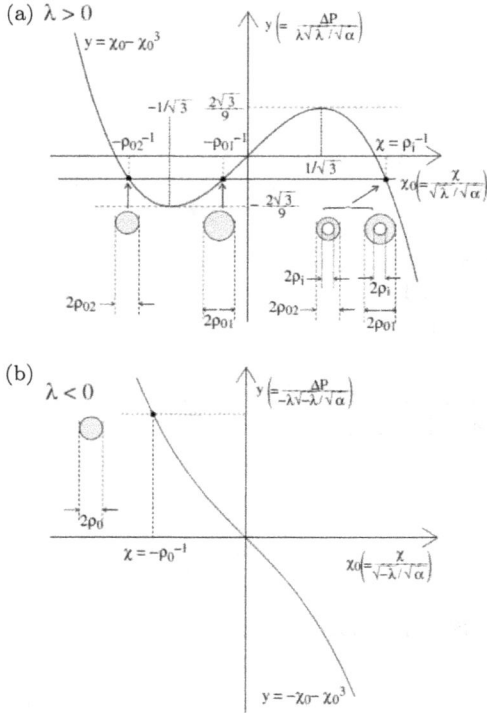

Fig. 5.5:   Illustration of equilibrium circle shape equation of $\Delta P(\kappa) = \lambda \kappa - \alpha \kappa^3$ (a) for $\lambda > 0$, and (b) for $\lambda < 0$. In practical calculations, $\Delta P_0 = \kappa_0 - \kappa_0^3$ and $\Delta P_0 = -\kappa_0 - \kappa_0^3$, i.e., (a) $\Delta P_0 = \Delta P/(\lambda\sqrt{\lambda}/\alpha)$ and $\kappa_0 = \kappa/(\sqrt{\lambda}/\sqrt{\alpha})$ and (b) $\Delta P_0 = (-\lambda\sqrt{-\lambda}/\alpha)$ and $\kappa_0 = \kappa/(\sqrt{-\lambda}/\sqrt{\alpha})$.

Source: Reprinted from Ref. [38], *Phys. Rev. Lett.*, 93(20), 206101, 2004.

solutions are divided into three cases:

(i) Two *circles* and two *tori* of different sizes coexist for $\lambda > 0$ and
   $0 > \Delta P > = \frac{2\sqrt{3}}{9} \frac{\lambda\sqrt{\lambda}}{\sqrt{\alpha}}$.

(ii) Only one *circle* exists for $\Delta P > 0$.

(iii) No compact circular domains can exist for $\lambda < 0$ for $\Delta P < 0$.

Case (ii) can be easily understood in mechanics: A negative tension and the elastic force ($\alpha > 0$) favor increasing the circle size, whereas a positive ΔP favors decreasing it; therefore, there exists an equilibrium size at which two tendencies balance. In case (iii) three forces are all outward; hence, they cannot balance each other. With such ideas, case (i) describes the solid domains grown in liquid phase and is then understandable: e.g., at inner circle of the equilibrium torus, both tension and elastic forces point to solid phase, on the other hand, the pressure points to fluid phase because $\Delta P < 0$ (in physics, $\Delta p = -g_0$, this is like the ice formation at temperature of 0°C). Actually, two *torus*-shaped domains of different sizes are shown in Fig. 10 in Ref. [18], and predicted two *circles* of different sizes are found in Fig. 4 in Ref. [20]. Note that the coexistence of two *tori* and two *circles* have not been predicted by the analysis by McConnell's approach (see Section 5.2.4), and only one equilibrium *circle* with a radius of $\frac{e^3 \Delta}{8} e^{\frac{\lambda}{\mu^2}}$ has been predicted. In other words, it is difficult to find the solutions of the two tori and two circles by using the McConnell's approach.

It is instructive here to note that for the cases of circular domains satisfying Eq. (5.38), the shape equation, Eq. (5.38), is simplified as follows:

$$k \frac{2\pi r_0}{\Delta} + r = \log \left| \frac{2\pi r_0}{\Delta} \right|, \tag{5.39}$$

with $k = \left(\frac{2\Delta P}{\mu_\perp^2} + \frac{11}{3}\pi^3 \right) \frac{\Delta}{2\tau}$ and $r = \frac{2\lambda}{\mu_\perp^2} - \frac{11}{12}\pi^2 - 1$. r_0 is the radius of the circular domain. Equation (5.39) can be solved graphically. According to the values and signs of the two constants k and r, which are also the linear function of the pressure ΔP and line tension λ, the solutions are theoretically divided into four cases as shown in Fig. 5.6.

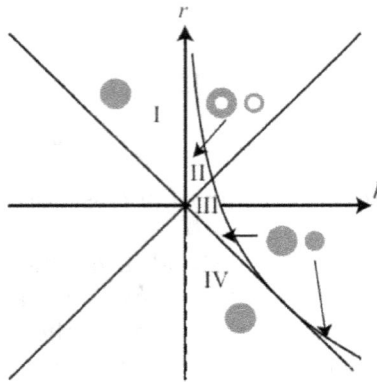

Fig. 5.6:   Morphological diagram of circular domains as evaluated by Eq. (5.34). The morphological diagram is divided into four regions: The shape equation has only one positive radius as the solution in region I ($r > -k$ and $k < 0$) and IV ($r < -k$ and $k > 0$). The solutions of this equation are two positive radii and one negative radius in region II $r < -\ln k - 1$ and $k > 0$) and two positive radii in region III ($r < -\ln k - 1$ and $-k < r < k$).

Source: Reprinted from Ref. [39], *J. Chem. Phys.*, 125, 224701, 2006.

Equation (5.39) has only one solution that is a positive radius for $k < 0$ and $r > -k$ (region I) and $k > 0$ and $r < -k$ (region IV). This equation has two positive solutions and one negative solution in $-\ln k - 1 > r$ and $0 < k < r$ (region II) and two positive solutions in $-\ln k - 1 > r$ and $-k < r < k$ (region III). This figure also includes the solution of Fig. 5.5, and the creation of circular domains in Fig. 5.6 is explained in the same way as in Fig. 5.5.

5.4.3.2.   *Quasi-polygon domains*

The quasi-polygon domain is seen as a branching phenomenon of the instability of a circle of radius ρ_0. We here consider a slightly distorted circle $\rho = \rho_0 + \sum_m b_m \exp(im\phi)$, where $i \cdot i = -1$, $0 \leq \phi \leq 2\pi$, $m = 0, \pm 1, \ldots, \pm\infty$, and $b_m^* = b_{-m}$ (see 5.2.4). The variations of domain area, boundary length, and the curvature elastic energy can be straightly obtained by using the variation method. The detail of derivation process is shown in the next Section 5.5.2.2. The calculation results are summarized as follows:

$$\delta \oint ds = \pi\rho_0 \left[2(b_0/\rho_0) + \sum_{m=0}^{\infty} m^2 \left| \frac{b_m}{\rho_0} \right|^2 \right], \tag{5.40}$$

$$\delta A = \pi \rho_0^2 \left[2(b_0/\rho_0) + \sum_{m=0}^{\infty} \left| \frac{b_m}{\rho_0} \right|^2 \right], \tag{5.41}$$

$$\delta \oint \kappa^2 ds = \pi \rho_0^{-1} \left[-2(b_0/\rho_0) + \sum_{m=0}^{\infty} |(2m^4 - 5m^2 + 2) \left| \frac{b_m}{\rho_0} \right|^2 \right]. \tag{5.42}$$

We here consider the case of $\delta A = 0$ and $\lambda < 0$, i.e., to maintain domain area constant, thus we obtain the linear term $(b_0/\rho_0) = -\frac{1}{2} \sum_m |b_m/\rho_0|^2$ (see also Eq. (5.20)). Substituting Eq. (5.40) and Eq. (5.42) into Eq. (5.37) yields the deformation energy

$$\delta F = \pi \sum_{m=0}^{\infty} [\lambda \rho_0 + \alpha \rho_0^{-1}(2m^2 - 1)](m^2 - 1) \left| \frac{b_m}{\rho_0} \right|^2. \tag{5.43}$$

The trivial case is $m = 1$, characterized by $\delta F = 0$. This means a translation of the circle. The m-th harmonic deformation happens when the coefficient of $|b_m/\rho_0|$ in Eq. (5.43) becomes negative, i.e., the domain increases to the condition of $\rho_0 > \rho_m$ with $\rho_m = \sqrt{(2m^2 - 1)\alpha/(-\lambda)}$, $(m \geq 2)$ [38]. The simple and analytic expression of $\rho_m/\rho_2 = \sqrt{2m^2 - 1}/\sqrt{7}$ agree well with the *ab initio* calculation results directly from Eq. (5.25) and Eq. (5.30) and the experimental measurements (Fig. 7 in [21]) performed by McConnell's group.

5.5. Shape Equation of Domains with *in-plane* Electric Dipoles [39]

5.5.1. *Derivation of shape equation in case with in-plane electric dipoles*

As mentioned in Section 5.3 (see also Chapter 4), microscope observation of DPPC monolayers shows *in-pane* orientational order, suggesting the presence of *in-plane* electric dipoles in one direction in domains. Because tilting the normal electric dipoles in one planner direction is the most possible picture of *in-pane* orientational order with *in-plane* electric dipoles. To study this effect on the shape of domains, we need to extend the free energy given by Eq. (5.25) to the

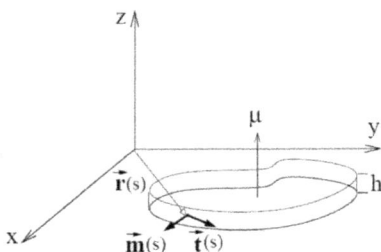

Fig. 5.7: Schematic illustration of the shape of a monolayer domain, whose dipoles μ uniformly tilt along the x direction with the same tilted angle θ. The geometric quantities describing its boundary curve are also shown, and h is the thickness of the monolayer (reprinted from Fig. 1 in Ref. [39]).

case with dipolar energy caused by *in-plane* electric dipoles. That is, the 2D domain energy F of Eq. (5.25) is replaced by [39]

$$F = \Delta P \int dA + \lambda \oint dl + F_\perp + F_\parallel, \qquad (5.44)$$

with additional dipolar energy $F_\parallel$ given by

$$F_\parallel = \frac{\mu_\parallel^2}{2} \oint \oint \frac{1}{r} dy dy' = \frac{\mu_\parallel^2}{2} \oint \oint \frac{(\vec{t}(l) \cdot \hat{y_0})(\vec{t}(s) \cdot \hat{y_0})}{|\vec{r}(l) - \vec{r}(s)|} dl ds, \qquad (5.45)$$

Here $\mu = (\mu_\perp, 0, \mu_\parallel) = \mu_0(\sin\theta, 0, \cos\theta)$ is postulated, and all dipoles are tilting along the x-direction with the same tilt angle θ (see Fig. 5.7).

The derivation process of shape equation for this case is the same as that shown in Section 5.4, but the calculation steps are lengthy and rather complex. The main calculation results are summarized as follows:

$$F_\parallel = \frac{\mu_\parallel^2}{2} \oint ds \int_h^L dx \left\{ \frac{1}{x} \sin^2 \phi(s) - \frac{1}{2}\kappa(s) \sin 2\phi(s) \right.$$

$$- \left[\frac{1}{4}\kappa_s(s) \sin 2\phi(s) + \frac{11}{24}\kappa(s) \sin^2 \phi(s) \right] x + o(x^2) \left. \right\}$$

$$\approx \frac{\mu_\parallel^2}{2} \ln \frac{L}{h} \oint \sin^2 \phi(s) ds$$

$$- \frac{1}{192}\mu_\parallel^2 L^2 \oint [11 + 13\cos 2\phi(s)]\kappa(s)^2 ds. \qquad (5.46)$$

$$\delta^{(1)} F_{\parallel} = \frac{\mu_{\parallel}^2}{2} \left(\frac{1}{L} \oint \sin^2 \phi(s) ds - \frac{11}{48} L \oint \kappa(s)^2 ds \right.$$

$$\left. - \frac{13}{48} L \oint \cos 2\phi \kappa(s)^2 ds \right) \oint \delta^{(1)} ds$$

$$- \frac{12}{96} \mu_{\parallel}^2 L^2 \oint \delta^{(1)} [\kappa(s)^2 ds]$$

$$+ \frac{\mu_{\parallel}^2}{2} \ln \frac{L}{h} \oint \delta^{(1)} [\sin^2 \phi(s) ds]$$

$$+ \frac{13}{96} \mu_{\parallel}^2 L^2 \oint \delta^{(1)} [\sin^2 \phi \kappa(s)^2 ds]$$

$$= -\frac{\mu_{\parallel}^2}{2} \oint \left[\frac{1}{L} \oint \sin^2 \phi(x) dx - \frac{11}{48} L \oint \kappa(x)^2 dx \right.$$

$$\left. - \frac{13}{48} L \oint \cos 2\phi \kappa(x)^2 dx \frac{1 + 3\cos 2\phi(s)}{2} \ln \frac{L}{h} \right] \kappa \Gamma(s) ds$$

$$+ \mu_{\parallel}^2 L^2 \oint \frac{39 \cos 2\phi(s) - 11}{192} \kappa^3 \Gamma(s) ds$$

$$- \frac{13}{24} \mu_{\parallel}^2 L^2 \oint \sin 2\phi \kappa \kappa_s \Gamma(s) ds$$

$$- \mu_{\parallel}^2 L^2 \oint \frac{39 \cos 2\phi(s) + 11}{96} \kappa_{ss} \Gamma(s) ds. \tag{5.47}$$

Here, it should be $\delta^{(1)} F = 0$ for any infinitesimal function $\Gamma(s)$ (see Eq. (5.37)). Hence combining Eqs. (5.37) with (5.47), we finally get the general shape equation of an equilibrium domain as follows:

$$\Delta P - \Lambda \kappa + \alpha \kappa^3 + \beta \kappa_{ss} + \sigma \kappa \kappa_s = 0, \tag{5.48}$$

where definitions of the coefficients are

$$\Lambda = \lambda - \frac{\mu_{\perp}^2}{2} \ln \frac{Le}{h} + \frac{\mu_{\parallel}^2 (1 + 3\cos 2\phi)}{4} \ln \frac{L}{h} + \frac{11}{48} \mu_{\perp}^2 L \oint \kappa(s)^2 ds$$

$$+ \frac{\mu_{\parallel}^2}{2L} \oint \sin^2 \phi(s) ds - \frac{1}{96} \mu_{\parallel}^2 L \oint (11 + 13\cos 2\phi) \kappa(s)^2 ds,$$

$$\alpha = \frac{11}{96}\mu_\perp^2 L^2 + \mu_\parallel^2 L^2 \frac{39\cos 2\phi - 11}{192},$$

$$\beta = \frac{11}{48}\mu_\perp^2 L^2 - \mu_\parallel^2 L^2 \frac{11 + 13\cos 2\phi}{96},$$

and

$$\sigma = -\frac{13}{24}\mu_\parallel^2 L^2 \sin 2\phi, \tag{5.49}$$

respectively. Again, the coefficients are functions of geometric parameters of domain shapes. Obviously, compared to the specific shape equation of Eq. (5.38) ($\mu_\parallel = 0$), the Eq. (5.48) is more complex with the complicated coefficients of $\kappa(s)$ and its derivatives, and $\kappa\kappa_s$ is involved. Noteworthy that Eq. (5.48) returns to Eq. (5.38) if $\mu_\parallel = 0$.

5.5.2.   *Solution of general shape Eq. (5.48)*

In this section, a couple of solutions of Eq. (5.48) are shown.

5.5.2.1.   *Circular shape domains*

Obviously, the shape equation, Eq. (5.48), is invariant with respect to the exchange of $\phi \to -\phi$ and $s \to -s$. This means that its solution must be symmetric with respect to the tilted direction of the x-axis of the dipole moments. This is the origin of the domain shapes with C_2 symmetry observed not only in DPPC monolayer (the CLS and Y-shaped domains in Figs. 1 and 3 of Ref. [41], respectively) but also in an L-α-dimyristoyl phosphatidic acid (DMPA) monolayer (Fig. 3 in Ref. [42]). Before discussing the circular solutions of Eq. (5.48), we rewrite it into the following form:

$$\Delta P - \Lambda_1 \kappa + \alpha_1 \kappa^3 - \mu_\parallel^2 \cos(2\phi)\left[\frac{3}{4}\ln\left(\frac{L}{h}\right) - \frac{13}{64}L^2\kappa^3\right]$$

$$+\beta\kappa_{ss} + \kappa\kappa_s = 0, \tag{5.50}$$

where the coefficients

$$\Lambda_1 = -\frac{\mu_\perp^2}{2}\ln\frac{Le}{h} + \frac{\mu_\parallel^2}{4}\ln\frac{L}{h} + \frac{11}{48}\mu_\perp^2 L \oint \kappa(s)^2 ds$$

$$+\frac{\mu_\parallel^2}{2L}\oint \sin^2\phi(s)ds - \frac{1}{96}\mu_\parallel^2 L \oint (11 + 13\cos 2\phi)\kappa(s)^2 ds, \tag{5.51}$$

and

$$\alpha_1 = \frac{11}{96}\mu_\perp^2 L^2 - \frac{11}{192}\mu_\parallel^2 L^2, \tag{5.52}$$

are independent of local variable $\phi(s)$; both of them are only dependent on global shapes of the domains. They are related to Λ and α in Eq. (5.50) by

$$\Lambda = \Lambda_1 + \frac{3}{4}\mu_\parallel^2 \ln\frac{L}{h}\cos 2\phi, \tag{5.53}$$

and

$$\alpha = \alpha_1 + \frac{13}{64}L^2\mu_\parallel^2 \cos 2\phi. \tag{5.54}$$

Therefore, a circle is a solution of Eq. (5.50) if its radius ρ_0 satisfies

$$\Delta P + \Lambda_1 \rho_0^{-1} - \alpha_1 \rho_0^{-3} = 0, \tag{5.55}$$

and

$$\frac{3}{4}\ln\left(\frac{L}{h}\rho_0^{-1}\right) - \frac{13}{64}L^2\rho_0^{-3} = 0 \quad \text{with } L = 2\pi\rho_0, \tag{5.56}$$

simultaneously. From Eq. (5.56), we have

$$\rho_c = \frac{h}{2\pi}\exp\left(\frac{13\pi^2}{12}\right) = 7.0 \times 10^3 h. \tag{5.57}$$

In the same figure of the CLS domain (Fig. 1 in Ref. [42]) a circular domain of $\rho_0 \approx 8\mu m$ appears. According to Tanford's estimation [43] for a saturated hydrocarbon chain with n carbon atoms, $h \leq (0.154 + 0.1265n)$ nm, we have $h \leq 2$ nm and $\rho_0 \leq 13.3\,\mu m$. Thus Eq. (5.57) predicts an interesting fact that *lipid chain of nanometer length can form domains with micrometer size.*

5.5.2.2. *Pressure-induced instability of the circular domains*

McConnell [44] theoretically predicted and experimentally demonstrated that the equilibrium circular domains become unstable by either increasing their size at a fixed surface pressure or changing the surface pressure for a fixed domain size: the circular domains sharply transform to m-side *quasipolygons* (the mth distorted harmonic shapes) initially, and then continuously develop into n-branched patterns and take on symmetries that are lower than that of the early *quasipolygons* [13, 21]. Moreover, on compressing a monolayer of racemic DPPC, one observes that the circular domains grow in area and form clover-leaf shape (CLS) [41]. Compared to the initial sharp transitions, the latter transitions are very smooth. McConnell attributed this transition difference to the coupling between the DPPC dipolar tilt and the domain shapes and sizes [41]. We here focus on this pressure-induced instability of the circular domains.

To investigate the circle to CLS domain growth of the monolayer of racemic DPPC by compression, i.e., $\Delta P \to \Delta P + \delta p$, where ΔP is the equilibrium pressure at the circular solution ρ_0 of Eq. (5.48), we assume that the growing domain is still a solution of the equilibrium Eq. (5.48) (see Section 5.4.3). We describe the growing domain by a slightly distorted circle given by

$$\mathbf{r}(s) = \rho_0 \mathbf{m}_0(s) + \Gamma(s)\mathbf{m}_0(s), \tag{5.58}$$

with

$$\mathbf{m}_0(s) = (\sin\phi_0(s), -\cos\phi_0(s)), \phi_0 = s/\rho_0. \tag{5.59}$$

This is the same procedure we took in Section 5.3. We then have the following variation form of Eq. (5.48) corresponding to a small change of the pressure δp:

$$\frac{\delta\Lambda}{\rho_0} - \frac{\delta\alpha}{\rho_0^3} + \delta p - \Lambda\delta + \alpha\delta\kappa^3 + \beta\delta\kappa_{ss} + \sigma\delta(\kappa_s\kappa) = 0. \tag{5.60}$$

Here the terms of variations of β and σ are absent because κ_s and κ_{ss} vanish for a circle. Considering that Λ_1 and α_1 are independent of $\phi(s)$ [Eqs. (5.51) and (5.52)], and substituting Eqs. (D.12), (5.55), and (5.56), we have the linear approximate equation for Γ of the above equation as follows,

$$\delta p + \frac{a_1}{\rho_0^2}\Gamma + a_2\left(\frac{1}{\rho_0}\Gamma_s + \rho_0\Gamma_{sss}\right) + a_3\Gamma_{ss} + a_4\rho_0^2\Gamma_{ssss} = 0, \tag{5.61}$$

where $\Gamma_{sss} = \mathrm{d}^3\Gamma/\mathrm{d}s^3$, $\Gamma_{ssss} = \mathrm{d}^4\Gamma/\mathrm{d}s^4$, and definitions of the coefficients are

$$a_1 = \Delta P\rho_0 + \frac{11\pi^2}{24}(2\mu_\perp^2 - \mu_\parallel^2) + \frac{13\pi^2\mu_\parallel^2}{8}\cos 2\phi_0,$$

$$a_2 = \frac{13\pi^2}{6}\mu_\parallel^2 \sin 2\phi_0,$$

$$a_3 = \Delta P \rho_0 + \frac{11\pi^2}{12}(2\mu_\perp^2 - \mu_\parallel^2) + \frac{13\pi^2}{12}\mu_\parallel^2 \cos 2\phi_0,$$

and

$$a_4 = \frac{11\pi^2}{24}(2\mu_\perp^2 - \mu_\parallel^2) - \frac{13\pi^2}{24}\mu_\parallel^2 \cos 2\phi_0,$$

respectively.

We then insert into Eq. (5.61) the expansion $\Gamma(s)$ that can be expanded by

$$\Gamma(s) = \sum_{m=-\infty}^{\infty} C_m \exp[im\phi_0(s)], \qquad (5.62)$$

with $i \cdot i = -1$, $0 \le \phi \le 2\pi$, $m = 0, \pm 1, \ldots, \pm\infty$ and $C_m^* = C_{-m}$ to keep the function to be real. The following relations are obtained

$$\frac{a_1}{\rho_0^2}\Gamma = \sum_m \exp(im\phi_0)\left\{ C_m\left[\frac{\Delta P}{\rho_0} + \frac{11\pi^2}{24\rho_0^2}(2\mu_\perp^2 - \mu_\parallel^2)\right]\right.$$
$$\left. + C_{m-2}\frac{13\pi^2\mu_\parallel^2}{16\rho_0^2} + C_{m+2}\frac{13\pi^2\mu_\parallel^2}{16\rho_0^2}\right\},$$

$$\frac{a_2}{\rho_0}\Gamma_s = \sum_m \exp(im\phi_0)\left[C_{m-2}\frac{13\pi^2\mu_\parallel^2}{12} - C_{m+2}\frac{(m+2)}{\rho_0^2}\frac{13\pi^2\mu_\parallel^2}{12}\right]$$

$$a_2\Gamma_{sss} = -\sum_m \exp(im\phi_0)\left[C_{m-2}\frac{(m-2)^3}{\rho_0^2}\frac{13\pi^2\mu_\parallel^2}{12}\right.$$
$$\left. + C_{m+2}\frac{(m-2)^3}{\rho_0^2}\frac{13\pi^2\mu_\parallel^2}{12}\right]$$

$$a_3\Gamma_{ss} = -\sum_m \exp(im\phi_0)\left\{C_m\frac{m^2}{\rho_0^2}\left[\Delta P\rho_0 + \frac{11\pi^2}{12}(2\mu_\perp^2 - \mu_\parallel)\right]\right.$$
$$\left. - C_{m-2}\frac{(m-2)^2}{\rho_0^2}\frac{13\pi^2\mu_\parallel^2}{24} - C_{m+2}\frac{(m+2)^2}{\rho_0^2}\frac{13\pi^2\mu_\parallel^2}{24}\right\}.$$

$$a_4 \rho_0^2 \Gamma_{ssss} = \sum_m \exp(im\phi_0) \left[C_m \frac{m^4}{\rho_0^4} \frac{11\rho_0^2 \pi^2}{24} (2\mu_\perp^2 - \mu_\|) \right.$$

$$\left. - C_{m-2} \frac{(m-2)^4}{\rho_0^4} \frac{13\pi^2 \rho_0^2 \mu_\|^2}{48} - C_{m+2} \frac{(m+2)^4}{\rho_0^4} \frac{13\pi^2 \rho_0^2 \mu_\|^2}{48} \right].$$

$$(5.63)$$

Hence, we combine them according to C_{m-2}, C_m, and C_{m+2} and let coefficients of $\exp(im\phi_0)$ vanish, an important recursive relation is obtained:

$$(m^2 - 1)\{\mu_\|^2[(m+1)(m+3)C_{m+2} + (m-1)(m-3)C_{m-2}]$$

$$+ [a + b(1 - m^2)]C_m\} - d\delta_{m,0} = 0, \qquad (5.64)$$

where $a = \frac{48\Delta P \rho_0}{13\pi^2}$, $b = \frac{22}{13}(2\mu_\perp^2 - \mu_\|^2)$, $d = \frac{48\rho_0^2}{13\pi^2}\delta p$, and $\delta_{m,0} = 1$ when $m = 0$, 0 otherwise. The solutions of Eq. (5.64) can be derived into two cases:

(i) For $\mu_\| = 0$ (*without dipole tilt*).

Equation (5.60) predicates an m-th harmonic shape transition (i.e., $C_m \neq 0$) under the following critical surface pressure:

$$\Delta P = \frac{11\pi^2 \mu_\perp^2 (m^2 - 1)}{12\rho_0}. \qquad (5.65)$$

Such a pressure-induced m-side quasipolygon transition has been observed and schematized in Ref. [21] (Figs. 8–12 therein). For DPPC, $\rho_0 \sim 1\,\mu m$, $\mu_\perp \sim 15\,\text{D}/50\,\text{Å}^2$, at a rough estimate, Eq. (5.65) gives $\Delta P \sim (m^2 - 1)m\,\text{Nm}^{-1}$, just in a reasonable order as observation [21]. The case of $m = 1$ means only the trivial translation of the circle, and corresponding $\Delta P = 0$ revealing translation to happen at $\Pi = g_0$. Our BAM observation shows similar results as shown in Fig. 5.8, where domain shapes and orientational structure of phospholipid monolayers were investigated using Brewster Angle Microscopy coupled with MDC measurement. The domain shapes change step-wise with surface pressure.

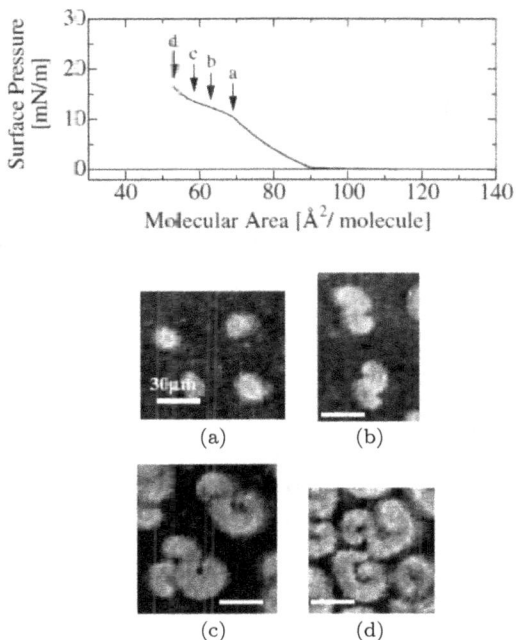

Fig. 5.8(a): L-DPPC surface pressure-area isotherms and BAM images of L-DPPC domain growth with a molecular compression rate of 0.62 Å^2/min: (a) 8.56, (b) 10.9, (c) 12.4, and (d) 13.9 mN/m. (a)–(d) in the graph corresponds to the images (scale bars represent 100 um) (reprinted from Ref. [45]).

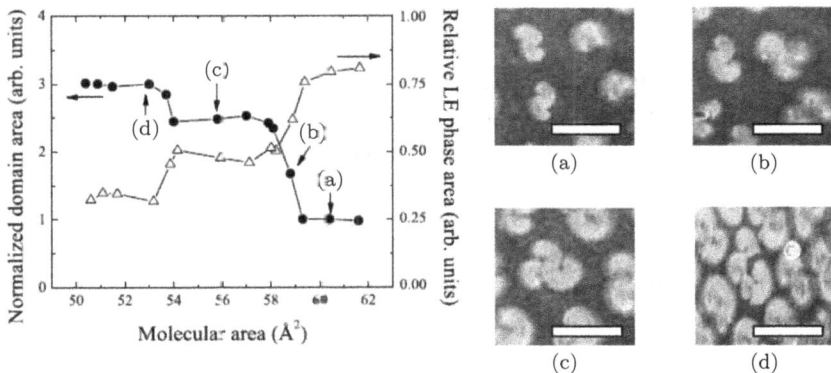

Fig. 5.8(b): Average domain area normalized in respect to smallest circular domain and relative total area of LE phase as a function of mean molecular area. Arrows denote positions of BAM images. Scale bars represent 50 um.

Source: From Ref. [17], *Jpn. J. Appl. Phys.*, 50, 051601, 2011.

(ii) For $\mu_{\parallel} \neq 0$ (*with dipole tilt*).

The dipolar tilt removes the above transition of eigen modes instead of the two collective deformation branches described as follows: Branch I happens at the beginning of the compression, i.e., $\delta p = 0$. From Eq. (5.64) with $|m| = 2n$ $(n = 0, \ldots \infty)$ and $C_0 = 0$ we find all of the even harmonic growing modes, $C_{|2n|}$, must vanish. The request of $C_0 = 0$ is consideration of the constraint of constant area of the domain at the beginning of the growth which leads into [46]

$$\frac{C_0}{\rho_0} = -\frac{1}{2} \sum_m \left| \frac{C_m}{\rho_0} \right|^2, \tag{5.66}$$

i.e., $C_0 = 0$ in the present linear approximation. But for odd modes the situation is quite different from the case of $\mu_{\parallel} = 0$ where only $C_{\pm 1} \neq 0$ can occur at $\Delta P = 0$ (see above). It is easy to find that when $|m| = 1$. Equation (5.64) allows not only $C_{\pm 1} \neq 0$ but also $C_{\pm 3} \neq 0$. If $C_{\pm 3} \neq 0$ happens, then all odd harmonic modes of C_{2n+1} are excited. Fitting to the whole CLS domain (Fig. 1 in Ref. [41]), we have the so-called translation (because $C_1 \neq 0$) induced growth solution

$$\rho = \rho_0 \left(1 + \frac{1}{5} \cos \Phi + \frac{8}{34} \cos 3\Phi \right), \tag{5.67}$$

see Fig. 5.9(a). Here $\Phi = \phi_0 - \pi/2$ is the azimuth of $\mathbf{m}_0(s)$. In the above equation the higher m-th harmonic modes $(m = 5, 7, \ldots)$ are neglected by considering them decaying rapidly with rates of m^{-2} at a magic tilted angle of the dipoles (i.e., $b = 0$; see further for details). The tilted dipole force can automatically excite the growth of $C_3 \neq 0$, which explains the domain formation with a three-arm labyrinth popularly observed in the 2D lipid monolayers [13, 41]. Branch II occurs at $\delta p \neq 0$ upon further compression. First, the identity of Eq. (5.64) at $|m| = 1$ still holds, therefore, $C_{\pm 1} \neq 0$ can exist, but from the experimental observation we assumed $C_{\pm 3} = 0$. From Eq. (5.64) with $|m| = 2n$ we have

$$C_{|2|} = -d^*/6,$$

$$C_{|4|} = (a^* - 3b^*)d^*/90,$$

$$C_{|6|} = [45 - (a^* - 3b^*)(a^* - 15b^*)]d^*/3150, \tag{5.68}$$

where $a^* = a/\mu_{\parallel}^2$, $b^* = b/\mu_{\parallel}^2$ and $d^* = d/\mu_{\parallel}^2$.

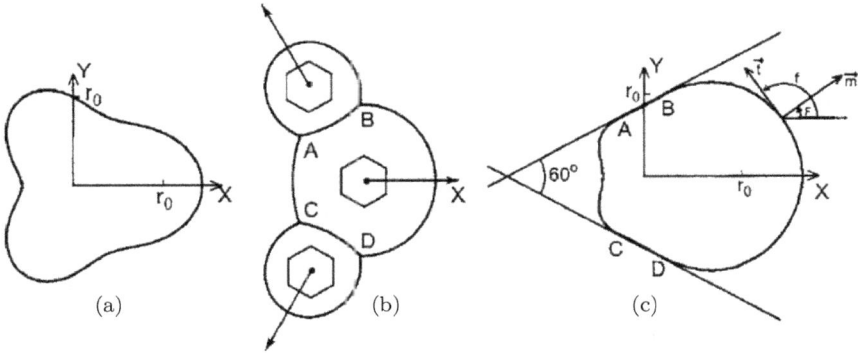

Fig. 5.9: Calculation of equilibrium shapes grown from a circle to (a) a $C2$-symmetric triangle [Eq. (5.67)] by a tilted dipole force along the x-axis at the beginning of a compression and then to a *tri*-domain structure illustrated in (b) with the increasing of the compression, where the main domain, the distorted hexagon, is also an equilibrium shape predicted by theory [Eq. (5.67)]. In practical computation, (a) is obtained by simulating a solid domain observed in a racemic monolayer of DPPC (Fig. 1 in Ref. [13], *J. Phys. Chem.*, 92, 5233, 1988) and (c) by simulating its main domain. The latter makes us find the magic dipolar tilt and its physical meaning, a zero shape formation energy [Eq. (5.67)].

A straightforward simulation with the observed main domain of the CLS *tri*-domain structure in racemic DPPA (Figs. 1 and 3(b) in Ref. [41]) gives such a pressure-induced solution

$$\rho = \rho_0(1+0.6\cos\Phi+0.06\cos2\Phi-0.06\cos4\Phi+0.02\cos6\Phi), \quad (5.69)$$

as shown in Fig. 5.9(c). Figure 5.9(c) reveals the two normal of the two straight sides (AB, CD) of the calculated main domain subtending exactly 120°. Both straight sides can serve as the platforms for the growth of the two side domains (Fig. 5.9(b)), both of which can be obviously described as solutions similar to Eq. (5.69). Substituting the above equation into Eq. (5.64), we found $b^* = 1/36$ and the tilted angle

$$\theta = \arctan\left(\frac{\mu_\|}{\mu_\perp}\right) \approx \arctan\sqrt{2} = 54.7°. \quad (5.70)$$

To understand this magic angle, we calculated the whole dipolar energy from Eqs. (5.29) and (5.46) for a circle of radius ρ as follows:

$$F_e = F_\perp + F_\| = (2\mu_-^2 - \mu_\|^2)\left(\frac{11\pi^2}{48} - \frac{1}{4}\ln\frac{2\pi\rho}{h}\right)2\pi\rho. \quad (5.71)$$

It becomes clear that the magic tilt makes a circular domain grown from $\rho = 0$ to ρ_0 with $F_e = 0$. In physics, the shape Eq. (5.48) reflects the mechanics equilibrium between two phases, while the zero shape formation energy describes the energy equilibrium of the solid domain grown from or melted into the fluid phase. In fact, the shape given in Fig. 5.9(c) is quite popular, e.g., in DMPA monolayer. Miller and Möhwald observed many such domains (Fig. 3 in Ref. [42]). A more direct evidence on the magic tilt may refer to the earlier X-ray diffraction studies in eicosanoic acid monolayers by Durbin *et al.* [47]. They found an $\sim 60°$ tilt in the nearest-neighbor NN phase. Similarly, MDC-SHG experiment on cyanobiphenyl liquid crystal monolayers by the author's group showed an $\sim 57°$ tilt in the condensed phase [48]. Given these, the CLS *tri*-domain structure observed in Ref. [41] can be viewed to form in two steps: the compression and the magic dipolar tilt first induce a translation $C_1 \neq 0$ and $C_3 \neq 0$ growth (a distorted triangle given in Fig. 5.9(a)) and the increasing surface pressure p then excites a deformation of all even eigenmodes and causes the domain splitting into three equilibrium distorted hexagons characterized by Eq. (5.69). With the domain fission, the dipoles in the two side domains also change their orientation as predicted by the authors of Ref. [41].

5.6. Exact Solution of Shape Equation

It is quite a task to solve the derived shape equations generally, but we can find some exact solutions. In this section, we focus on such solutions.

5.6.1. *Numerical calculation of exact solution*

It is instructive here to show some exact solutions, though both Eqs. (5.38) and (5.48) are highly nonlinear and difficult to solve analytically. As we assume a circular domain as the most possible solution and discuss the stable and unstable conditions between the radius of a circle and the pressure ΔP and line tension in Section 5.4. Here, we will show that Eq. (5.38) has an exact solution if $\Delta P = 0$. In physics, the vanishing ΔP means that the solid and fluid phases of the lipid monolayer coexist and the π_c-plateau is present in the $\pi - A$ isotherm [20, 50].

Substituting geometrical relations of $\kappa_s = d\kappa/ds = -(1/2)d\kappa^2/d$ and $\kappa_{ss} = (1/2)\kappa d^2\kappa^2/d\phi^2$ into Eq. (5.38), and multiplying both sides of the equation with $d\kappa^2/d\phi$, we obtain its first integral

$$\left(\frac{d\kappa^2}{d\phi}\right)^2 = \left(\frac{2\Lambda}{\alpha}\right)\kappa^2 - \kappa^4 + C, \tag{5.72}$$

where C is a constant of integration. For a domain, $\mathbf{r}(s)$ is a finite closed curve and κ may take extreme values at some points at which $d\kappa^2/d\phi \propto \kappa_s = 0$. By choosing $s = 0$ at a certain extreme point and letting $\phi(0) = \pi/2$, we have the apparent solution of

$$\kappa_s = -d\phi/ds = -a\sqrt{1 + b\sin\phi(s)}, \tag{5.73}$$

where $a = \sqrt{\Lambda_0/\alpha_0}$, $b = (\kappa_0/a)^2 - 1$ and $\kappa_0 = \kappa(0)$.

Obviously, if $b = 0$, then the above equation gives a circle with radius a. Moreover, Eq. (5.73) can describe both round shapes with convex or concave cusp. As illustrated, Fig. 5.8 shows the numerical results of the equation with $a = 1$, $b = \pm 0 : 7$, and $\pm\phi(0)$. Here, we have chosen $s = 0$ at the cusp of the boundary and $m(0)$ along y. We see that $\phi(0)$ with opposite signs can yield very distinct domain shapes (the upper and lower curves in the figure). These shapes were indeed observed at the π_c-plateau in the *dipamitoyl-phosphatidylcholine* (DPPC) monolayer [32], and the upper shapes in Fig. 5.10 were called boojum-like therein. Hence, the solution (5.73) is not only exact but also satisfactorily predicts the interesting shapes observed in the experiment.

5.6.2. *Analytical solutions to the shape equation*

In the previous section, the solution of shapes of boojum-like domains was numerically illustrated. Here we show more exactly the rigorous solutions of (5.38) described by using elliptic functions.

5.6.2.1. $\Delta P = 0$

We restart from Eq. (5.38), and instead of Eq. (5.72), we rewrite the solution in the way as

$$\sin\phi(s) = \frac{\kappa^2(s) - \Lambda/\alpha}{\kappa^2(0) - \Lambda/\alpha}. \tag{5.74}$$

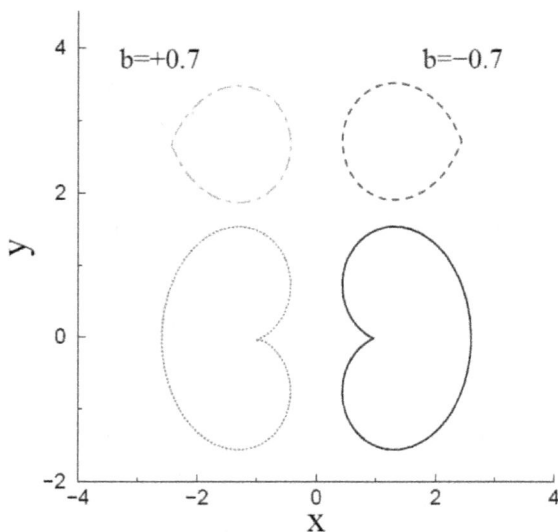

Fig. 5.10: Illustration of boojum- (upper) and kidney-like (low) shapes calculation by Eq. (5.72) with $a = 1$, $b = \pm 0 : 7$ and $\pm \emptyset(0)$. The different sign of b gives
the left- and right- "kidneys" and "boojums", respectively.
Source: Reprinted from Ref. [54], *Int. J. Modern Phys.* B, 22(13), 2047–2053,
2008.

Substituting $\sin \phi = 1 - 2 \sin 2\overline{\phi}$, $\overline{\phi} = \phi/2 - \pi/4$ and using the
relation $\kappa(s) = -d\phi/ds$, we finally have the solution of $\phi(s)$ as

$$\frac{1}{2}\kappa(0)s = \int_0^{\overline{\phi}} \frac{d\overline{\phi}}{\sqrt{1 - k^2 \sin^2 \overline{\phi}}} = F(\overline{\phi}, k). \tag{5.75}$$

Here $k = \sqrt{2[1 - \Lambda/\alpha\kappa^2(0)]}$, and $F(\overline{\phi}, k)$ is the first kind elliptic integral with modulus k. With the definition of Jacobi's func-
tions [49], Eq. (5.75) is rewritten as $\sin \overline{\phi} = \mathrm{sn}[\kappa(0)s/2]$. Considering
$\cos \phi == -2 \sin \overline{\phi} \cos \overline{\phi}$, the analytical solution of $\mathbf{r}(s) = (x(s), y(s))$
is derived as follows,

$$\tilde{x}(\tilde{s}) = -2 \int_0^{\tilde{s}} \mathrm{sn}(s')\mathrm{cn}(s')ds'$$

$$\tilde{y}(\tilde{s}) = \int_0^{\tilde{s}} [1 - 2\mathrm{sn}^2(s')]ds', \tag{5.76}$$

Here x, y and s are expressed in dimensionless forms, respectively:
$\tilde{x} = \kappa(0)x/2$, $\tilde{y} = \kappa(0)y/2$, and $\tilde{s} = \kappa(0)s/2$. In other words,

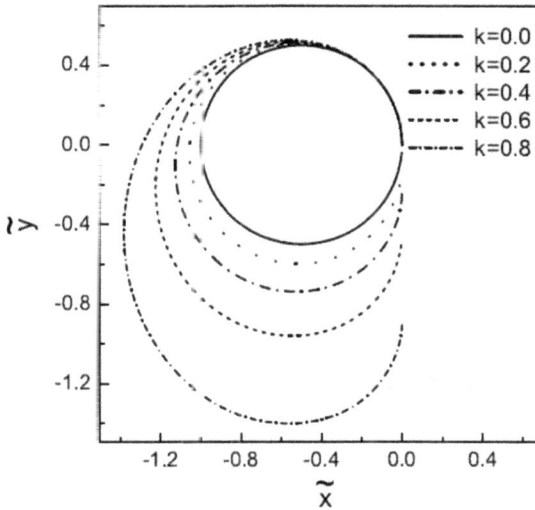

Fig. 5.11: Solution of 2D domain shape Eq. (5.8) at $\Delta P = 0$. Equation (5.72) shows a domain growth from an initial circle ($k = 0$) to dipping circle and bean-like shapes with increasing the modulus $k = \sqrt{2[1 - \lambda/\alpha\kappa^2(0)]}$ from 0 to 1. *Source*: Reprinted from Ref. [49]. *Eur. Phys. J. E*, 27, 81–86, 2008.

the domain shapes at $\Delta P = 0$ can be topologically described by Eq. (5.76) with the change of k from 0 to 1, according to the change of Λ.

Figure 5.11 shows the numerical results of the equation. We see that Eq. (5.76) indeed reveals a domain growth from a circle to bean-like forms from $k = 0$ to 1. It also agrees with the experimental observation for a monolayer of *dipalmitoyl-phosphatidylcholine* (DPPC), in which the domain growth from a small circle to a larger bean-like shape (Figs. 2(a), (b) and (c) in Ref. [52]) does happen at π_c-plateau of LE-LC coexistence with near the same π_c (3.8, 3.9, and 4.2 mN/m, respectively).

5.6.2.2. $\Delta P \neq 0$

Solving Eq. (5.38) becomes more complicated due to the presence of $\Delta P (\neq 0)$. We first multiply both sides of the shape equation with κ_s and get its first integral,

$$\Delta P\kappa - \frac{\Lambda}{2}\kappa^2 + \frac{\alpha}{4}\kappa^4 + \alpha\kappa_s^2 = C. \qquad (5.77)$$

By choosing $s = 0$ at which $\kappa_s = 0$ again and doing a transformation of $\kappa(s) - \kappa(0) = [c_1\eta(s) + c_2]^{-1}$ with appropriate coefficients c_1 and c_2, we reexpress the above equation as follows,

$$\frac{ds}{d\eta} = \frac{1}{\sqrt{4\eta^3 - g_2\eta - g_3}}, \tag{5.78}$$

where $g_2 = [\overline{\Lambda^2} + 6\overline{\Lambda}\kappa^2(0) - 3\kappa^4(0) - 12\Delta\bar{P}\kappa(0)]/48$,

$$g_3 = [27\Delta\overline{P^2} - 36\overline{\Lambda}\Delta\bar{P}\kappa(0) - 9\overline{\Lambda}\kappa^4(0) + 18\overline{\Lambda^2}\kappa^2(0) - \overline{\Lambda^3}]/1728,$$

$\Delta\bar{P} = \Delta P/\alpha$ and $\overline{\Lambda} = \Lambda/\alpha$.

The solution to Eq. (5.78) is just the Weierstrass elliptic function $\wp(s)$ [39]. Hence, we can find a general solution of Eq. (5.38) as

$$\kappa(s) - \kappa(0) = \frac{6(\overline{\Lambda}\kappa(0) - \Delta\bar{P} - \kappa^3(0))}{24\wp(s) + 3\kappa^2(0) - \overline{\Lambda}}. \tag{5.79}$$

Given these, it is the ease of finding shape solution of $\mathbf{r}(s)$ by $d\mathbf{r}(s)/ds = (\cos\phi, \sin\phi)$ and $d\phi/ds = -\kappa(s)$. In other words, one can always calculate the solution diagram of the domains vs. $\Delta\bar{P}$ and $\overline{\Lambda}$, though it is a lengthy and huge numerical process.

5.6.2.3. *Cusp*

A cusp appears in the boundary of a peach-like domain (see Fig. 4 in Ref. [53]). This feature is included in the solution of Eq. (5.79). Using a linear instability analysis, it can be proved that above the following critical pressure given by Eq. (5.65), a circle of radius ρ_0 is deformed into a m-side *quasipolygon* for $m \geq 2$. For the case of $m = 1$ or $\Delta P = 0$, an infinitesimal deformation means a trivial translation of the circle, and a finite deformation is revealed and shown in Fig. 5.12. Thus, it is expected that the remainder case of $m = 0$ corresponds with a peach-like deformation with a cusp.

To prove this conjecture, a small change in the pressure around the critical surface pressure ΔP_0 is considered. It will be convenient to introduce several dimensionless quantities for discussion: $\Delta\tilde{P} = \Delta\bar{P}\rho_0^3$ and $\tilde{\Lambda} = \overline{\Lambda}\rho_0^2$, $\tilde{s} = s/\rho_0$, $\tilde{\kappa}(\tilde{s}) = \rho_0\kappa(\tilde{s}\rho_0)$; here $\kappa(0) = -1/\rho_0$ is assumed. Because the initial equilibrium domain is a circle with radius ρ_0, according to the definition of Eq. (5.65), we immediately

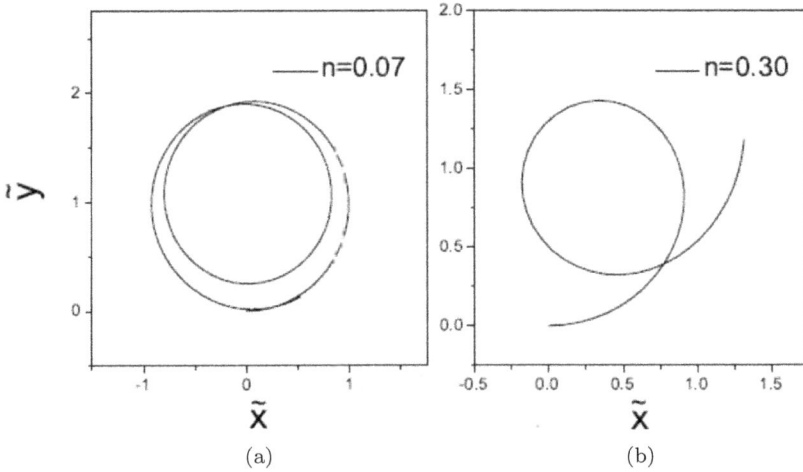

Fig. 5.12:   Domain shapes calculated with Eq. (10) for $\lambda = 3$, $\Delta P = -2 + \delta p$, and $64/27 \geq \delta p \geq 0$, where Weiestrass elliptic function-associated polynome $f(\eta) = 4\eta^3 - g_2\eta - g_3$ has three distinct real roots. It shows a circle can deform in sequence into two ellipse-like domains where one has a cusp and the other has a dip (a), and one peach-like domain with a cusp (b).

Source: Reprinted from Ref. [49]. *Eur. Phys. J. E*, 27, 81–86, 2008.

have $\Delta \tilde{P}_0 = -2$, $\tilde{\Lambda} = 3$ and $\tilde{\kappa}(0) = -1$. Let the surface pressure be $\Delta \tilde{P} = \Delta \tilde{P}_0 + \delta p$, Eq. (5.79) is then

$$\tilde{\kappa}(\tilde{s}) = -1 - \frac{\delta p}{4\wp(\tilde{s})}, \qquad (5.80)$$

where the $\wp(s)$ is associated with $\tilde{g}_2 = g_2\rho_0^4 = \delta p/4$ and $\tilde{g}_3 = g_3\rho_0^6 = (\delta p/8)^2$. In practical calculations, the Weiestrass elliptic function $\wp(s)$ has to be converted to the forms in common use, e.g., the Jacobi's elliptic functions. If the three roots of $f(\eta) = 4\eta^3 - \tilde{g}_2\eta - \tilde{g}_3$ are distinct and real, then $\wp(\tilde{s})$ can be written as [51]

$$\wp(\tilde{s}) = e_3 + \frac{e_1 - e_3}{sn^2(\sqrt{e_1 - e_3}\tilde{s})}, \qquad (5.81)$$

with modulus $k = \sqrt{(e_2 - e_3)/(e_1 - e_3)}$, under assumption $e_1 > e_2 > e_3$. In this case, the associated determinant $G = \tilde{g}_2^3 - 27\tilde{g}_3^2 > 0$ results in $0 \leq \delta p \leq 64/27$, and the three roots are respectively $e_1 = n\cos\theta$, $e_2 = n\cos(\theta + 4\pi/3)$, and $e_3 = n\cos(\theta + 2\pi/3)$, with $n = \sqrt{\delta p/12}$ and

$\cos 3\theta = 9n/4$. Figure 5.12(a) shows the numerical results calculated from Eqs. (5.80) and (5.81) for the two n values, 0.07 and 0.30.

We see that with the increase of δp the circle deforms in sequence into two elliptic domains in which one has a cusp and the other has a "dip" at the boundary crossing their short symmetric axis, and one peach-like domain with a cusp. But if the determinant G is negative or $\delta p < 0$ and $\delta p > 64/27$, Eq. (5.57) cannot be used in that the two roots of $f(\eta)$ are complex. In this situation, we find that it is still possible to convert the $\wp(s)$ into Jacobi's elliptic functions. We first write the three roots as $e_1 = a + ib$, $e_2 = -2a$, and $e_3 = a - ib$, respectively, where $i \cdot i = -1$, $a = -(s+t)/2$, $b = \frac{\sqrt{3}}{2}(s - t)$, and $s = (\tilde{g}_3/8 + \sqrt{r})^{1/3}$, $t = \tilde{g}_2/(12s)$, $r = (-\tilde{g}_2/12)^3 + (-\tilde{g}_3/8)^2$. After some mathematic manipulations (the details see the Appendix), we finally get the equation of $\wp(\tilde{s})$ as

$$\wp(\tilde{s}) = 2c^2 \frac{[1 + \text{cn}(2c\tilde{s})]}{\text{sn}^2(2c\tilde{s})} - 2a - c^2, \qquad (5.82)$$

where $c^2 = \sqrt{9a^2 + b^2}$ and the moduli for both sn and cn are $\sqrt{1/2 + 3a/2c^2}$. Figure 5.13 shows the numerical results.

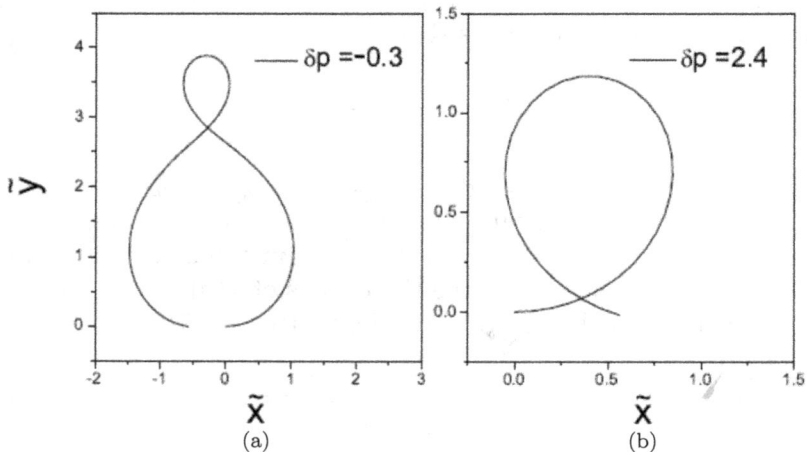

Fig. 5.13: When $\delta p < 0$ and $\delta p > 64/27$ the Weierstrass elliptic function-associated $f(\eta)$ has complex roots, the circle deforms into the peach-like shapes with one cusp at boundary regardless of δp being negative (a) or positive (b).
Source: Reprinted from Ref. [49], *Eur. Phys. J. E*, 27, 81–86, 2008.

We see that the solution of Eq. (5.80) for $f(\eta)$ having complex roots gives ellipse-like shapes with one cusp at the boundary crossing their long symmetric axis regardless of $\delta p < 0$ and $\delta p > 64/27$; The only difference is that there are two such peach-like domains for given one negative δp while only one for $\delta p > 64/27$. The theoretical domains displayed in Figs. 3 and 4 are nicely in agreement with the experiments: the ellipses appear in Fig. 5 in Ref. [23] and Fig. 4 in Ref. [11], and the ellipse-like domains with one cusp at the boundary are mostly reported in all the mentioned literature [20, 50, 53].

5.7. Summary

In this chapter, the mathematical method "differential geometry" is introduced to describe 1D curves on 2D flat surfaces. Electrostatic dipolar energy arising from dipole–dipole interactions is expanded by Taylor expansion, and the shape equation describing domain shapes was obtained in an approximate form. The derived equation is very similar to the Helfrich shape equation, and suggesting that the double line integral of the dipolar energy can be approximately expressed as a sum of an additionally negative line tension and a curvature-elastic energy of the domain boundary. By solving the derived shape equation, a variety of domains of shapes, e.g., circular shapes of solid domain surrounded by fluid phase, Clover-leaf shapes (CLS), Kidney- and boojum-like domains that abound in a lipid monolayer, deformed shapes of their transitions to torus, D form, S form, serpentine manner form, m-sided quasi-polygon form, and others, were found to be obtained. However, we are still not clear on the physics of the generation of additional negative line tension and curvature-elastic energy from dipolar energy, in terms of the electrostatic Maxwell stress. In the next chapter, we further discuss the details.

References

[1] H. M. McConnell and V. T. Moy, Shapes of finite two-dimensional lipid domains, *J. Phys. Chem.*, 92, 4520–4525 (1988).

[2] K. Terasawa, *Introduction to Mathematics* (Sugaku Gairon), Chapter 2, Iwanami, 1960 (in Japanese).

[3] Z. C. Ou-Yang, J. X. Liu, and Y. Z. Xie, *Geometric Methods in the Elastic Theory of Membranes in Liquid Crystal Phases*, World Scientific, Singapore, 1999.

[4] I. Langmuir, The constitution and fundamental properties of solids and liquids II. Liquids, *J. Am. Chem. Soc.*, 39, 1848 (1917).

[5] G. L. Gaines, *Insoluble Monolayers at the Liquid-Gas Interface*, Wiley-Interscience, New York, 1966.

[6] V. M. Kaganer, H. Möwald, and P. Dutta, Structure and phase transitions in Langmuir monolayers, *Rev. Mod. Phys.*, 71, 779 (1999).

[7] M. Losche, E. Sackmann, and H. Möwald, A fluorescence microscopic study concerning the phase-diagram of phospholipids, *Ber. Bunsenges. Phys. Chem.*, 87, 848 (1983).

[8] R. Peters and K. Beck, Translational diffusion in phospholipid monolayers measured by fluorescence microphotolysis, *Proc. Natl Acad. Sci. USA*, 80, 7183 (1983).

[9] V. von Tscharner and H. M. McConnell, An alternative view of phospholipid phase behavior at the air-water interface. Microscope and film balance studies, *Biophys. J.*, 36, 409 (1981).

[10] D. Hönig and D. Möbius, Direct visualization of monolayers at the air–water interface by Brewster angle microscopy, *J. Phys. Chem.*, 95, 4590– 4592 (1991).

[11] S. Henon and J. Meunier, Microscope at the Brewster angle: Direct observation of first-order phase transitions in monolayers, *Rev. Sci. Instrum.*, 62(4), 936 (1991).

[12] R. M. Weis and H. M. McConnell, Two-dimensional chiral crystals of phospholipid, *Nature*, 310, 47–49 (1984).

[13] V. T. Moy, D. J. Keller, and H. M. McConnell, Molecular order in finite two-dimensional crystals of lipid at the air–water interface, *J. Phys. Chem.*, 92, 5233–5238 (1988).

[14] T. Yamamoto, *Dielectric and Geometrical Properties of Dipolar Monolayers in Liquid Crystal Phase*, Dr. Thesis, Tokyo Institute of Technology, September 2007.

[15] P. Krüger and M. Lösche, Molecular chirality and domain shapes in lipid monolayers on aqueous surfaces, *Phys. Rev. E*, 62, 7031 (2000).

[16] T. Yamamoto, T. Aida, T. Manaka, and M. Iwamoto, Chiral phase separation of a monolayer domain comprised of racemic mixture of chiral phospho-lipids due to the electrostatic energy, *Colloids and Surfaces A: Physicochem. Eng. Aspects*, 321, 151–157 (2008).

[17] M. Weis, W. Ou-Yang, T. Yamamoto, Y. Matsuoka, T. Manaka, and M. Iwamoto, Observation of continuous and quantized domain size and shape evolution in monolayers at air–water interface, *Jpn. J. Appl. Phys.*, 50(5), 051601 (2011).

[18] R. M. Weis and H. M. McConnell, Cholesterol stabilizes the crystal-liquid interface in phospholipid monolayers, *J. Phys. Chem.*, 89, 4453–4459 (1985).

[19] H. E. Gaub, V. T. Moy, and H. M. McConnell, Reversible formation of plastic two-dimensional lipid crystals, *J. Phys. Chem.*, 1986, 90(8), 1721–1725.

[20] K. J. Stine and D. T. Stratmann, Fluorescence microscopy study of Langmuir monolayers of stearylamine, *Langmuir*, 8, 2509–2514 (1992).

[21] K. Y. C. Lee and H. M. McConnell, Quantized symmetry of liquid monolayer domains, *J. Phys. Chem.*, 97, 9532–9539 (1993).

[22] D. Andelmann, F. Brochard, and J. F. Joanny, Phase transitions in Langmuir monolayers of polar molecules, *J. Chem. Phys.*, 86, 3673–3681 (1987).

[23] D. J. Keller, J. P. Korb, and H. M. McConnell, Theory of shape transitions in two-dimensional phospholipid domains, *J. Phys. Chem.*, 91, 6417–6422 (1987).

[24] H. M. McConnell and V. T. Moy, Shapes of finite two-dimensional lipid domains, *J. Phys. Chem.*, 92, 4520–4525 (1988).

[25] Samuel A. Safran, *Statistical Thermodynamics of Surfaces, Interfaces, and Membranes* (Frontiers in Physics), CRC Press (January 29, 2003).

[26] J. M. Deutch and F. E. Low, Theory of shape transitions of two-dimensional domains, *J. Phys. Chem.*, 96, 7097–7101 (1992).

[27] J. A. Miranda, Closed form results for shape transitions in lipid mono-layer domains, *J. Phys. Chem. B*, 103, 1303–1307 (1999).

[28] J. Plateau, *Statique Experimentable et Theoretique des Liquides Soumis aux Seules Forces Moleculaires*, Gauthier Villars, Paris, 1873.

[29] T. Young, III, An essay on the cohesion of fluids, *Philos. Trans. R. Soc. London*, 95, 65 (1805).

[30] P. S. Laplace, *Traite de Mecanique Celeste*, Gauthier Villars, Paris, 1839.

[31] G. Wulff, On the question of speed of growth and dissolution of crystal surfaces, *Z. Kristallogr.* 34, 449–530 (1901).

[32] W. Helfrich, Elastic properties of lipid bilayers: Theory and possible experiments, *Z. Naturforsch. C*, 28(11), 693–703 (1973).

[33] Z.-C. Ou-Yang and W. Helfrich, Instability and deformation of a spherical vesicle by pressure, *Phys. Rev. Lett.*, 59, 2486 (1987).

[34] H. Naito, M. Okuda, and Z.-C. Ou-Yang, Preferred equilibrium structures of a smectic-A phase grown from an isotropic phase: Origin of focal conic domains, *Phys. Rev. E*, 52, 2095 (1995).

[35] Z. C. Ou-Yang, Anchor ring-vesicle membranes, *Phys. Rev. A*, 41, 4517 (1990).

[36] M. Mutz and D. Bensimon, Observation of toroidal vesicles, *Phys. Rev. A*, 43, 4525 (1991).

[37] A. S. Rudolph, B. R. Ratna, and B. Kan, Self-assembling phospho-lipid filaments, *Nature*, 352, 52–55 (1991).

[38] M. Iwamoto and Z.-C. Ou-Yang, Shape deformation and circle insta-bility in two-dimensional lipid domains by dipolar force: A shape- and size-dependent line tension model, *Phys. Rev. Lett.*, 93(20), 206101 (2004). (The term $(2m^2 - 3)$ in Eq. (30) should be corrected as $(2m^2 - 1)$).

[39] M. Iwamoto, F. Liu, and Z. C. Ou-Yang, Shape and stability of two-dimensional lipid domains with dipole–dipole interactions, *J. Chem. Phys.*, 125, 224701 (2006).

[40] S. A. Langer, R. E. Goldstein, and D. P. Jackson, Dynamics of labyrinthine pattern formation in magnetic fluids, *Phys. Rev. A*, 46, 4894 (1992).

[41] V. T. Moy, D. J. Keller, H. E. Gaub, and H. M. McConnell, Long-range molecular orientational order in monolayer solid domains of phospho-lipid, *J. Phys. Chem.*, 90, 3198–3202 (1986).

[42] A. Miller and H. Möwald, Collecting two-dimensional phospholipid crystals in inhomogeneous electric fields, *Europhys. Lett.*, 2, 67 (1986).

[43] C. Tanford, *The Hydrophobic Effect* Wiley, NY, 1973, 1980.

[44] H. M. McConnell, Harmonic shape transitions in lipid monolayer domains, *J. Phys. Chem.*, 94, 4728–4731 (1990).

[45] T. Aida, T. Yamamoto, W. Ou-Yang, T. Manaka, and M. Iwamoto, Study of domain shapes and orientational structure of phospholipid monolayers using Maxwell displacement current and Brewster angle microscopy, *Jpn. J. Appl. Phys.*, 47, 411 (2008).

[46] Z.-C. Ou-Yang and W. Helfrich, Bending energy of vesicle mem-branes: General expressions for the first, second, and third variation of the shape energy and applications to spheres and cylinder, *Phys. Rev. A*, 39, 5280 (1989).

[47] M. K. Durbin, A. Malik, A. G. Richter, R. Ghaskadvi, T. Got, and P. Dutta, Transitions to a new chiral phase in a Langmuir monolayer, *J. Chem. Phys.*, 106, 8216 (1997).

[48] A. Tojima, T. Manaka, M. Iwamoto, and Z.-C. Ou-Yang, Detection of phase transition of monolayers at the air–water interface by compres-sion using Maxwell displacement current and optical second harmonic generation, *J. Chem. Phys.*, 118(12), 5640–5649 (2003).

[49] M. Iwamoto, F. Liu, and Z.-C. Ou-Yang, Elliptic function solutions for the shapes of lipid monolayer domains, *Euro. Phys. J. E.*, 27(1), 81–86 (2008).

[50] K. Kjaer and J. A. Nielsen, Ordering in lipid monolayers studied by synchrotron X-ray diffraction and fluorescence microscopy, *Phys. Rev. Lett.*, 58, 2224 (1987).

[51] H. Hancok, *Lecture on the Theory of Elliptic Functions*, Dover, New York, 1958.

[52] C. W. McConlogue and T. K. Vanderlick, A close look at domain formation in DPPC monolayers, *Langmuir*, 13, 7158 (1997).

[53] D. K. Schwartz, M. W. Tsao, C. M. Knobler, and J. C. Knobler, Domain morphology in a two-dimensional anisotropic mesophase: Cusps and boojum textures in a Langmuir monolayer, *J. Chem. Phys.*, 101, 8258 (1994).

[54] M. Iwamoto, F. Liu, and Z.-C. OuYang, Domain shapes in lipid monolayers studies as polar cholesteric liquid crystals, *Int. J. Mod. Phys. B.*, 22(13), 2047–2053 (2008).

Chapter 6

Differential Geometry Method for Analyzing Domain Shapes II

This chapter serves as a companion to Chapter 5, which employed the differential geometry method for analyzing domain shapes. In Chapter 5, it was demonstrated that the double line integral of the dipolar energy could be represented as a sum of an additionally negative line tension and a curvature-elastic energy of the domain boundary using the Taylor approximation method. However, the understanding of the physics behind this double integral in terms of electrostatics and dielectric physics was deemed insufficient. This chapter aims to address this gap by illustrating that the double line integral of the dipolar energy, arising from dipole–dipole interactions, is the origin of electrostatic Maxwell stress. In simpler terms, electrostatic Maxwell stress acts as a bridge between electric dipolar polarization energy and mechanical force, influencing a variety of domain shapes. This chapter focuses on the inner texture of domains in monolayers, modeling ordered dipolar orientations while considering the liquid crystalline property. It explores the effects of electrostatic Maxwell stress on domain shapes, assuming that polarizations in domains appear parallel in the direction along directors. Using boojum domain shapes observed in fatty acid monolayers as an example, the chapter delves into the treatment of these shapes, starting from circular domains, and assumes that electrostatic Maxwell stress functions as a perturbation force to circular domains. Paying attention to inner texture of domains in monolayers, ordered dipolar orientations are modeled with consideration of the liquid crystalline property, and the effects of electrostatic Maxwell stress on domain shapes are treated. In the modeling, polarizations in domains are assumed to appear parallel in the direction along directors. As an example, boojum domain shapes visualized in fatty acid monolayers are treated starting from circular domains, assuming electrostatic Maxwell stress works as perturbation force to circular domains.

6.1.   Liquid Crystalline Nature of Domains

6.1.1.   *Observation of liquid crystalline nature of domains in monolayers*

As discussed in Chapter 5, electrostatic dipolar energy arising from dipole–dipole interactions plays an important role in the formation of domain shapes in monolayers. The basic approach proposed by McConnel is very elegant, but it is no longer sufficient to explain the presence of a variety of domain shapes, possibly owing to the mathematical difficulty in treating dipolar energy terms represented using double integrals. For example, clover shaped domain shapes having straight lobes in monolayers was predicted by McConnel's approach, but this approach seems insufficient to show visualized clover-shaped domains [1].

Polarized fluorescence microscopes (PFM) and polarized Brewster angle microscopes (pBAM) using polarized incident lights clearly show that molecules in domains in DPPC monolayers exhibit local *in-plane* orientational ordering of constituent molecules (where its range is longer than several micrometers) and that local average *in-plane* orientations are non-uniformly distributed [2]. Further, PFM and pBAM images suggest that the positions showing the same local average orientations in DPPC monolayers follow the curvatures of the bent lobes of these domains [3, 4]. These experimental facts suggest an idea that *in-plane* orientational ordering of constituent molecules is involved in the bending shape lobes of the clover-shaped domains in lipid monolayers. Interestingly, the situation that local *in-plane* average orientations are distributed non-uniformly in domains is analogous to liquid crystals, where their average orientational distributions are deformed non-uniformly by boundary conditions. Moreover, grazing X-ray diffraction studies have shown that the structures of the condensed phase of fatty acid monolayers are like reminiscent of hexatic-phases of liquid crystals, see also Chapter 4 (Section 4.1). These facts suggest that monolayers preserve liquid-crystalline natures. Accordingly, Rudnick and Bruinsma discussed the *in-plane* orientational distributions and shapes of domains in fatty acid monolayers by employing the continuum theory of liquid crystals [5–7]. A fatty acid molecule has one (hydrophobic) alkyl chain that is linearly connected with one hydrophilic hydroxide head, and is an achiral (non-chiral) molecule. Fatty acid monolayers stabilize domains of condensed phase, in which fatty acid molecules show local *in-plane* orientational order. These domains show non-uniform patterns of *in-plane* orientations that are called "virtual" boojum, where "virtual" implies that defects directly contributing to creating the patterns of *in-plane* orientations are excluded from the domains (virtual defects) (see Fig. 6.1).

6.1.2.   *Free energy of liquid crystalline monolayers and domain shapes*

In this section, we approach the geometry of domain shapes in monolayers using a different method from that employed in Chapter 5. We emphasize the liquid crystalline nature of monolayers, particularly by delving into the inner texture of domains. Subsequently, we

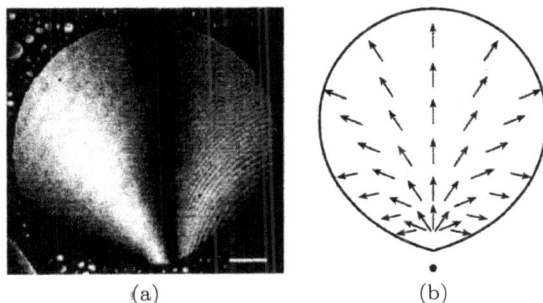

(a)   (b)

Fig. 6.1:   Brewster angle micrograph of a domain in monolayers of palmitic (hexadecanoic) acids (a) [Reproduced from Ref. [8]]. The radius of this domain is 196 μm. Local *in-plane* orientations in this domain show a so-called *virtual boojum texture* that is an orientational distribution generated by a "virtual" defect (b) [Reproduced from Ref. [6]]. The black dot that is at the outside of the domain in (b) is a virtual defect.

consider the distinctive polar structure of amphiphilic monolayers. Through this approach, we demonstrate that electrostatic Maxwell stress serves as a bridge between electric dipolar polarization energy and mechanical force, leading to a variety of domain shapes.

6.1.2.1.   *Free energy of monolayers based on Frank elastic energy theory*

The non-uniform in-plane orientations of domains in monolayers bear an analogy to the nature of liquid crystals. Consequently, it is instructive to employ the Frank elastic energy theory for the analysis of the inner texture of domains. Rudnick and Bruinsma described the free energy of a domain in monolayers in the form [7]

$$F_{\text{RB}} = \int dS \left[\frac{1}{2} k_s (\boldsymbol{\nabla}_\parallel \cdot \mathbf{c}(\mathbf{R}))^2 + \frac{1}{2} k_b (\boldsymbol{\nabla}_\parallel \times \mathbf{c}(\mathbf{R}))^2 \right] + \oint ds \lambda(\gamma).$$

(6.1)

The first term is 2D Frank elastic energy that is generated by the orientational deformation of unit vectors $\mathbf{c}(\mathbf{R})$ to the direction of local *in-plane* orientational order (*in-plane* directors). The second term is line energy arising from anisotropic line tension $\lambda(\gamma)$ that is a function of the angle γ between the normal $\mathbf{m}(s)$ of the domain boundary and *in-plane* director $\mathbf{c}(\mathbf{r}(s))$, $\cos \gamma = \mathbf{m}(s) \cdot \mathbf{c}(\mathbf{r}(s))$. k_s and k_b are elastic constants of monolayers with respect to *in-plane* splay $\boldsymbol{\nabla}_\parallel \cdot \mathbf{c}(\mathbf{R})$ and bend $\boldsymbol{\nabla}_\parallel \times \mathbf{c}(\mathbf{R})$ deformations. dS is the area element and $\mathbf{R}$ is the positional vector to the interior of the domain. ds is the line element and $\mathbf{r}(s)$ is the positional vector to the boundary line of the domain.

Rudnick and Bruinsma [7] adopted the one-constant approximation $k_s = k_b$, and expanded anisotropic line tension in a Fourier series

$$\lambda(\gamma) = \lambda_0 + a_1 \cos \gamma + a_2 \cos 2\gamma,$$

(6.2)

where higher order terms are neglected. Equation (6.2) can be viewed as the so-called "anchoring energy" of liquid crystals; this represents the extent that the orientation of liquid crystals is bounded by the boundary that fixes the orientations of directors [6]. Indeed, because of the fact that fatty acid monolayers on water surface show non-centrosymmetric orientational structures, spontaneous splay s_0

contributes to the Frank elastic energy of domains in monolayers, where these contributions have the form

$$-k_s s_0 \int dS \nabla_{\parallel} \cdot \mathbf{c}(\mathbf{R}). \tag{6.3}$$

By using divergence theorem (Gauss' theorem) for 2D flat surfaces, Eq. (6.3) is rewritten in the form

$$-k_s s_0 \oint ds\, \mathbf{c}(\mathbf{r}(s)) \cdot \mathbf{m}(s) = -k_s s_0 \oint ds \cos\gamma. \tag{6.4}$$

This implies that the second term of Eq. (6.2) corresponds to the elastic energy arising from spontaneous splay and characterizes domains that show non-centrosymmetric orientational structures on water surface [6–8]. The third term of Eq. (6.2) is the 2D analogue of the anchoring energy of nematic liquid crystals.

Minimizing the free energy F_{RB} given by Eq. (6.1) with respect to *in-plane* directors leads to torque balance equations (that ensure the balance of torques that are applied to *in-plane* directors)

$$\nabla^2 \Theta = 0, \tag{6.5}$$

and

$$k\mathbf{m}(s) \cdot \nabla \mathbf{c}(\mathbf{R})\|_{\mathbf{R} \to \mathbf{r}(s)} = \lambda'(\theta - \Theta)\|_{\mathbf{R} \to \mathbf{r}(s)}, \tag{6.6}$$

in the interior and boundary of the domain, respectively. Here $k = k_s = k_b$ (one-constant approximation), $\lambda'(\gamma)$ is the first derivative of $\lambda(\gamma)$ with respect to γ, and $\Theta(\mathbf{R})$ is the angle between *in-plane* director and the reference axis, $\mathbf{c}(\mathbf{R}) = (\cos\Theta, \sin\Theta)$. Here a Lagrange multiplier ΔP is used to ensure that the area of the domain is constant, and the term $\Delta P\, dS$ is added to Eq. (6.1).

Minimizing Eq. (6.1) with the added term $\Delta P dS$ with respect to the shapes of the domain leads to a relation that is satisfied at equilibrium.

$$\lambda_0 \kappa = \Delta P + F(s), \tag{6.7}$$

with

$$F(s) = \frac{1}{2}k(\nabla_{\parallel}\Theta)^2 + \lambda_1 \nabla_{\parallel} \cdot \mathbf{c}(\mathbf{r}(s))$$
$$+ \lambda_2[3\kappa \cos 2\gamma + 4\cos\gamma \nabla_{\parallel} \cdot \mathbf{c}(\mathbf{r}(s))]$$
$$+ 4\mathbf{c}(\mathbf{r}(s)) \cdot \mathbf{t}(s)(\mathbf{t}(s) \cdot \nabla_{\parallel}\mathbf{c}(\mathbf{r}(s))) \cdot \mathbf{m}(s). \tag{6.8}$$

Note that the original paper by Rudnick and Bruinsma uses somewhat different parametrization [7] that leads to a somewhat different equation, but their derived equation is equivalent to Eq. (6.7).

6.1.2.2. *Boojum texture of in-plane directors in domains of fatty acids*

Domains in fatty acids show virtual boojum texture of *in-plane* directors. The *in-plane* directors of virtual boojum that is generated by a virtual defect located at $(x_d, 0)$ has the form

$$\mathbf{c}(\mathbf{R}) = (\cos 2\phi_d, \sin 2\phi_d), \tag{6.9}$$

with

$$\cos \phi_d = \frac{x - x_d}{r_d}, \tag{6.10}$$

$$\sin \phi_d = \frac{y}{r_d}, \tag{6.11}$$

and

$$r_d = \sqrt{(x - x_d)^2 + y^2}. \tag{6.12}$$

Here r_d is the distance between a position $\mathbf{R} = (x, y)$ and the position of virtual defect $(x_d, 0)$. Equation (6.9) is a solution of Eq. (6.5).

We derive the solution of Eqs. (6.6) and (6.7) for a circular domain that has the form

$$\mathbf{r}(s) = r_0(\cos \theta, \sin \theta), \tag{6.13}$$

where r_0 is the radius of this domain and the center of this domain is located at the origin. *In-plane* directors are uniform $\mathbf{c} = (1, 0)$ in the domain for $x_d \to \infty$, and are strongly deformed as the virtual defect approaches the domain boundary $x_d \to r_0$, see Eq. (6.9). Indeed, for the case of $\lambda_2 = 0$, substituting Eqs. (6.13) and (6.9) leads to x_d in the form

$$\frac{x_d}{r_0} = \frac{k}{\lambda_1 r_0} + \sqrt{1 + \frac{k}{\lambda_1 r_0}}. \tag{6.14}$$

Equation (6.14) implies that the position of the virtual defect is determined by a single parameter $k/\lambda_1 r_0$ that represents the relative magnitudes of 2D Frank elastic energy and anisotropic line tension.

For the case that $k/\lambda_1 r_0$ is large, the position of virtual defects is far from the domains because 2D Frank elastic energy suppresses the deformations of *in-plane* directors. For the case that $k/\lambda_1 r_0$ is small, the position of virtual defects is close to the domain boundary because anisotropic line tension tends to direct *in-plane* directors in normal to the domain boundary.

We now consider the contribution of the third term of Eq. (6.2), $\lambda_2 \neq 0$, as a small perturbation. We derive the forms of *in-plane* directors and the shapes of domains by the first-order term with respect to λ_2. This calculation shows that domains stabilize cusp when the deformations of *in-plane* deformations are large enough because these domains can decrease the free energy by deforming their shapes so that the domain boundary is (more) parallel to *in-plane* directors; the virtual boojum and shapes of domains in fatty acid monolayers are due to the competition between 2D Frank elastic energy and anisotropic line tension.

Rudnick and Bruinsma argued that the third term of Eq. (6.2) is necessary for circular domains to show cusps [7]. By contrast, Rivière and Meunier argued that circular domains show cusps when λ_2 is zero, meanwhile Frank elastic constants for splay and bend are different, $k_s \neq k_t$ [8]. They experimentally extracted the position of the virtual defect from pBAM micrographs as a function of the radius of domains and performed a curve fitting with a theory for $\lambda_2 = 0$ and $k_s \neq k_b$. Interestingly, they obtained the best fit for the cases that the Frank elastic constants for bend deformations are almost zero but they also implied that experimental errors are relatively large ($0.3 < (k_s - k_b)/(k_s + k_b) < 1$ and $2.5 \times 10^4 \, \text{m}^{-1} < 2\lambda_1/(k_s + k_b) < 3.6 \times 10^4 \, \text{m}^{-1}$) [8]. Finally, Petty and Lubensky argued that circular domains show cusps even for $\lambda_2 = 0$ and $k_s = k_b$, but λ_1 is relatively large that anisotropic line tensions $\lambda(\gamma)$ can be locally negative [9].

These competing arguments suggest that employing the Frank elastic energy theory is helpful to predict cusps which are observed from circular domains of fatty acid monolayers, but the physical mechanism that stabilizes these cusps is not conclusive. This could be attributed to the fact that the elastic free energy, as defined in Eq. (6.1), is constructed exclusively based on the system's symmetry, with a specific emphasis on the liquid crystalline nature of domains. This situation encourages a more in-depth exploration of the physics behind 2D Frank elastic energy and anisotropic line tensions, where the exploration is conducted with consideration of distinctive polar property of the constituent amphiphilic molecules in monolayers, specifically considering the electric dipolar energies discussed in Chapter 5.

6.2.   Electrostatic Mechanisms of 2D Frank Elastic Energy and Anisotropic Line Tensions

In this section, significant attention is given to the electrostatic energy resulting from dipole–dipole interactions within liquid crystalline domains. In Chapter 5, it was shown that electrostatic energy due to dipole–dipole interactions plays an important role in the formation of domain shapes in monolayers. The fact that condensed phase domains in monolayers show *in-plane* orientational ordering implies that these domains have *in-plane* spontaneous polarizations, where their directions are tangent to the monolayer planes, and electrostatic interactions between *in-plane* dipoles spontaneously can store electrostatic energy in these domains. In greater detail, monolayers exhibit a unique characteristic in which they possess spontaneous polarization arising from the orientational alignment of amphiphilic molecules on the water surface. Consequently, it is straightforward to deduce that the storage of electrostatic dipolar energy is concomitant with the *in-plane* orientational ordering of the constituent rod-like molecules within the domains of monolayers.

As detailed in Chapter 5, McConnell and Moy investigated the impact of electrostatic energy resulting from in-plane dipole–dipole interactions on the shapes of domains in monolayers, and predicted that electrostatic energy stored by dipole–dipole interactions contributes to the elongation of domains. However, their study was confined to scenarios where in-plane dipoles are uniform across these domains [10]. In Chapter 5, with consideration of this electrostatic energy due to interactions among dipoles, shape equations expressed using geometric parameters were derived by means of differential geometry method [11, 12]. The derivation revealed that the double line integral of the dipolar energy can be expressed as a sum, incorporating an additionally negative line tension and a curvature-elastic energy. Nevertheless, the physical mechanism governing non-uniform distributions of local in-plane orientations in domains still requires discussion from the perspective of electrostatics in dielectric physics (see Chapter 2).

In the following sections, while considering the distinctive polar structure of amphiphilic monolayers, we assume that polarizations manifest in the direction along *directors*, akin to the presentation in Chapter 5. This assumption, as we demonstrate, leads to the understanding that the physical significance of the double integral of dipolar energy is tied to electrostatic Maxwell stress discussed in Chapters 1 and 2.

While addressing electrostatic energy resulting from dipole–dipole interactions, it is instructive to revisit the definition of polarization. The area densities of electric dipoles, denoted as $\mu_\perp$ in Eq. (5.7), represent spontaneous polarizations $P_{0\perp}$. It is important to note that polarizations are essentially electric dipoles per unit volume for bulk materials (refer to Chapter 2), and for surface polarizations, they are electric dipoles per unit area. Considering this definition of polarization in dielectric physics, domain shapes exhibiting a liquid-crystalline nature, as represented using directors, can be modeled by accounting for electrostatic energies arising from dipole–dipole interactions. This modeling involves assuming that dipolar polarization occurs parallelly in the direction along directors.

6.2.1. *Electrostatic energy generated from non-uniform spontaneous polarizations*

For the case of domains that show non-uniform spontaneous polarizations, electrostatic energy arising from the generation of these spontaneous polarizations has the form

$$F_{\text{ele}} = \frac{1}{2} \int dS_i \int dS_j \left[\frac{\mathbf{P_0}(\mathbf{R}_i) \cdot \mathbf{P_0}(\mathbf{R}_j)}{|\mathbf{R}_i - \mathbf{R}_j|^3} \right. $$
$$\left. - 3 \frac{\mathbf{P_0}(\mathbf{R}_i) \cdot (\mathbf{R}_i - \mathbf{R}_j) \mathbf{P_0}(\mathbf{R}_j) \cdot (\mathbf{R}_i - \mathbf{R}_j)}{|\mathbf{R}_i - \mathbf{R}_j|^5} \right]. \quad (6.15)$$

$\mathbf{R}i$ and $\mathbf{R}j$ represent positional vectors within the interior of the domain, and dS_i and $d\bar{S}_j$ denote area elements at $\mathbf{R}i$ and $\mathbf{R}j$, respectively (with $\mathbf{R}i$ and $\mathbf{R}j$ independently spanning the interior of the domain). As $\mathbf{R}i$ and $\mathbf{R}j$ are 2D vectors tangent to the monolayer plane, contributions to electrostatic energy arising from normal and in-plane spontaneous polarizations do not couple. Our attention is solely directed toward the electrostatic energy, $F_{\|}$, arising from *in-plane* spontaneous polarization.

Using a mathematical formula

$$
\nabla_j \frac{\mathbf{R}_i - \mathbf{R}_j}{|\mathbf{R}_i - \mathbf{R}_j|} = -\frac{\pi}{\Delta} \mathbf{I}\delta(\mathbf{R}_i - \mathbf{R}_j) - \frac{\mathbf{I}}{|\mathbf{R}_i - \mathbf{R}_j|^3}
$$
$$
+ 3\frac{(\mathbf{R}_i - \mathbf{R}_j)(\mathbf{R}_i - \mathbf{R}_j)}{|\mathbf{R}_i - \mathbf{R}_j|^5}, \tag{6.16}
$$

to the electrostatic energy due to *in-plane* dipole–dipole interactions, the following relation is obtained in the form

$$
F_{\|} = \int dS \frac{\pi}{\Delta} P_{0\|}^2(\mathbf{R})
$$
$$
- \frac{1}{2} \int dS_i \int dS_j P_{0\|}(\mathbf{R}_j) \cdot \nabla_j \left(\frac{P_{0\|}(\mathbf{R}_i) \cdot (\mathbf{R}_i - \mathbf{R}_j)}{|\mathbf{R}_i - \mathbf{R}_j|^3} \right). \tag{6.17}
$$

Here, $\mathbf{I}$ ($\equiv e_x e_x + e_y e_y$) is the identical operator with respect to 2D spaces. We here use ∇ as 2D gradient that is parallel to the monolayer plane to simplify the notation (where, thus far, we have used $\nabla_{\|}$ for this gradient).

With a mathematical vector formula

$$
\mathbf{n} \cdot \boldsymbol{\nabla}(V(\mathbf{R})\mathbf{n} \times \mathbf{A}(\mathbf{R})) = \mathbf{A}(\mathbf{R}) \cdot \nabla V(\mathbf{R}) + V(\mathbf{R})\nabla \cdot \mathbf{A}(\mathbf{R}). \tag{6.18}
$$

Equation (6.17) is rewritten in the form

$$
F_{\|} = -\frac{1}{2} \int dS_i \oint dS_j \frac{P_{0\|}(\mathbf{r}(S_j)) \cdot \mathbf{m}(S_j)P_{0\|}(\mathbf{R}_i) \cdot (\mathbf{R}_i - \mathbf{r}(S_j))}{|\mathbf{R}_i - \mathbf{r}(S_j)|^3}
$$
$$
+ \frac{1}{2} \int dS_i \int dS_j \frac{\nabla_j \cdot P_{0\|}(\mathbf{R}_j)P_{0\|}(\mathbf{R}_i) \cdot (\mathbf{R}_i - \mathbf{R}_j)}{|\mathbf{R}_i - \mathbf{R}_j|^3}. \tag{6.19}
$$

Here the first term of Eq. (6.17) is discarded because this can be included in the Lagrange multiplier in the following analysis.

The integrands of both of the two terms in Eq. (6.19) are rewritten by using a relationship

$$\nabla_i \frac{1}{|\mathbf{R}_i - \mathbf{R}_j|} = -\frac{\mathbf{R}_i - \mathbf{R}_j}{|\mathbf{R}_i - \mathbf{R}_j|^3}. \tag{6.20}$$

Using Eq. (6.18) to the factors of the form $\mathbf{P}(\mathbf{R}_i) \cdot \nabla_i |\mathbf{R}_i - \mathbf{R}_j|^{-1}$ leads to electrostatic energy arising from *in-plane* spontaneous polarizations in the form

$$
\begin{aligned}
F_\| = &\frac{1}{2} \int dS_i \int dS_j \frac{\nabla_i \cdot \mathbf{P}_{0\|}(\mathbf{R}_i) \nabla_j \cdot \mathbf{P}_{0\|}(\mathbf{R}_j)}{|\mathbf{R}_i - \mathbf{R}_j|} \\
&- \int dS_i \oint dS_j \frac{\nabla_i \cdot \mathbf{P}_{0\|}(\mathbf{R}_i) \mathbf{P}_{0\|}(\mathbf{r}(S_j)) \cdot \mathbf{m}(S_j)}{|\mathbf{R}_i - \mathbf{r}(S_j)|} \\
&+ \frac{1}{2} \oint dS_i \oint dS_j \frac{\mathbf{P}_{0\|}(\mathbf{r}(S_i)) \cdot \mathbf{m}(S_i) \mathbf{P}_{0\|}(\mathbf{r}(S_j)) \cdot \mathbf{m}(S_j)}{|\mathbf{r}(S_i) - \mathbf{r}(S_j)|}.
\end{aligned}
\tag{6.21}
$$

This form indeed has the form of Eq. (2.31) (see Chapter 2) and thus equals to electrostatic energy arising from dielectric polarizations. Noteworthy that energy w_{pol} that is necessary to generate dielectric polarizations by external fields is zero, but electrostatic energy by spontaneous polarizations is generated in the absence of external fields.

6.2.2. *Induced charges by in-plane spontaneous polarizations [13]*

Equation (6.21) has a simple physical meaning: Induced charges that are generated by *in-plane* spontaneous polarizations at the boundary of domains and in the interior of these domains have the forms

$$\sigma_{\text{ind}}(s) = \mathbf{P}_{0\|}(\mathbf{r}(s)) \cdot m(s), \tag{6.22}$$

and

$$\rho_{\text{ind}}(\mathbf{R}) = -\nabla \cdot \mathbf{P}_{0\|}(\mathbf{R}), \tag{6.23}$$

respectively, see also Chapter 2 (Eqs. (2.17) and (2.18)). In the physics of liquid crystals, the divergence of directors that correspond to the divergence of *in-plane* spontaneous polarizations $\mathbf{P}_{0\parallel}$ in our model here, represents the splay deformations of these polarizations, see Section (4.4) in Chapter 4, and thus Eq. (6.23) implies that splay deformations of *in-plane* spontaneous polarizations generate induced charges at the interiors of domains and thus lead to increase electrostatic energy. In this way, we can bridge the dielectric physics of monolayers and the physics of liquid crystals of monolayers.

For the case of domains that show uniform *in-plane* spontaneous polarizations, $\nabla \cdot \mathbf{P}_{0\parallel}(\mathbf{R}) = 0$, these *in-plane* spontaneous polarizations generate induced charges at the boundaries of these domains, $\sigma_{\text{ind}}(s) \neq 0$, but not in their interiors, $\rho_{\text{ind}} = 0$. By contrast, for the case of domains that show non-uniform *in-plane* spontaneous polarizations, $\nabla \cdot \mathbf{P}_{0\parallel} \neq 0$ these *in-plane* spontaneous polarizations generate induced charges not only at the boundaries of these domains $\sigma_{\text{ind}}(s) \neq 0$ but also in their interiors $\rho_{\text{ind}} \neq 0$. This analysis implies that the electrostatic effects of non-uniform *in-plane* spontaneous polarizations can be effectively modeled by appropriately assuming the spatial distribution of electric induced charges within the interiors of domains.

6.2.3.   *Electrostatic energy of non-uniform spontaneous polarizations expressed by induced charges*

An advantage of employing Eq. (6.21) is its potential expansion as a series of the curvature (κ) of the contour line, utilizing the Frenet–Serret theorem, similar to the approach in Chapter 5. This is possible because the integrands of these energy contributions are proportional to $1/|\mathbf{R}_i - \mathbf{R}_j|$. Since 2D positional vectors $\mathbf{R}_i$ and $\mathbf{R}_j$ are involved, it is necessary to extend the method elucidated in Chapter 5 to 2D. Further details can be found in Refs. [14] and [15].

We here treat the case of domains, where the gradients of induced charges in their interiors are relatively small, and thus these induced charges are expanded in the form $\rho_{\text{ind}}(\mathbf{R}_j) = \rho_{\text{ind}}(\mathbf{R}_i) + \delta\rho_{\text{ind}}(\mathbf{R}_i, \mathbf{R}_j)$ with $\delta\rho_{\text{ind}}(\mathbf{R}_i, \mathbf{R}_j) \ll \rho_{\text{ind}}(\mathbf{R}_i)$ for sufficiently small distance between $\mathbf{R}_i$ and $\mathbf{R}_j$. In this case, the first term $F_\parallel$ of Eq. (6.21) is expanded in the form

$$\frac{1}{2} \int dSL K_{\text{ele}}(\mathbf{R})(\boldsymbol{\nabla} \cdot \mathbf{P}_{0\parallel}(\mathbf{R}))^2$$

$$- \int dSL K_{\text{ele}}(\mathbf{R}) s_{0f}(\mathbf{R}) \boldsymbol{\nabla} \cdot \mathbf{P}_{0\parallel}(\mathbf{R}), \tag{6.24}$$

with

$$LK_{\text{ele}}(\mathbf{R}) = \int dS_j \frac{1}{|\mathbf{R} - \mathbf{R}_j|}, \tag{6.25}$$

$$LK_{\text{ele}}(\mathbf{R})s_{0f}(\mathbf{R}) = \frac{1}{2} \int dS_j \frac{\delta\rho_{\text{ind}}(\mathbf{R}, \mathbf{R}_j)}{|\mathbf{R} - \mathbf{R}_j|}. \tag{6.26}$$

Systematic expansions of Eqs. (6.25) and (6.26) with respect to curvatures are possible by using Frenet–Serret theorem [14,15]. The first and second terms of Eq. (6.24) have the forms of 2D Frank elastic energy with respect to splay deformations of in-plane spontaneous polarizations and their spontaneous splays, respectively, but $K_{\text{ele}}(\mathbf{R})$ and $s_{0f}(\mathbf{R})$ are, in general, functions of 2D positions $\mathbf{R}$ and reflect the shapes of the domain.

In a similar manner, the second term $F_{\parallel}$ of (6.21) is expanded in the form

$$\oint ds g(s) \mathbf{P}_{0\parallel}(\mathbf{r}(s)) \cdot \mathbf{m}(s), \tag{6.27}$$

with

$$g(s) = -\int dS_j \frac{\nabla_j \cdot \mathbf{P}_0(\mathbf{R}_j)}{|\mathbf{r}(s) - \mathbf{R}_j|}. \tag{6.28}$$

Indeed, $g(s)$ has a relationship with $K(R)$ and $s_{0f}(R)$, $g(s) = -LK(r(s))\nabla \cdot \mathbf{P}_{0\parallel}(\mathbf{R})|R \to r(s) + 2LK(r(s))s_{0f}(r(s))$, and thus is also systematically expanded with respect to the curvature of the domain boundary.

Finally, the third term $F_{\parallel}$ of Eq. (6.21) is expanded in the form

$$\oint ds \left[\frac{1}{2}\log\frac{L}{\Delta}\sigma_{\text{ind}}^2(s) + \frac{L^2}{96}\kappa^2(s)\sigma_{\text{ind}}^2(s) - \frac{1}{8}L^2\left(\frac{d}{ds}\sigma_{\text{ind}}(s)\right)^2 \right], \tag{6.29}$$

where a systematic expansion was performed in the manner same as in Section 4.3.

We here note that Eqs. (6.24), (6.27), and (6.29) are quadratic functions of $\rho_{\text{ind}}(R)$ and $\sigma_{\text{ind}}(s)$. This is because of the principle of superposition of electrostatics; electrostatic energy of the systems of many electric charges is the sum of electrostatic energy contributions arising from two-body Coulomb interactions. Equations (6.24), (6.27) and (6.29) are the first, second and third terms of $F_\parallel$, and they are functions of *in-plane* spontaneous polarizations.

In this way, the free energy used by Rudnick and Bruinsma, Eq. (6.1), represented as a function of *in-plane* director can be expressed as functions of electrostatic spontaneous polarizations, i.e., as functions of electrostatic charges, under the assumption that dielectric spontaneous polarization appears along the direction of directors in domains, according to the *in-plane* ordering.

6.2.4. *Relationship between electrostatic dipolar energy and elastic energy of domain in monolayers with spontaneous dipolar polarizations*

It is instructive here to treat monolayers composed of rod-shaped molecules with electric permanent dipoles at their center of gravity to relate *in-plane* directors and *in-plane* spontaneous polarizations. These electric dipoles are directed parallelly along the long-axes of the rod-shaped molecules, $\boldsymbol{\mu} = \mu(0, 0, 1)_m$, where the suffix "m" indicates the molecular coordinate system. The Euler rotational matrix (see Chapter 2) is used to represent this electric dipole in the monolayer coordinate system.

We here use the monolayer coordinate system, where the z-direction is normal to the monolayer and the x-direction is parallel to the *in-plane* director $\mathbf{c}$. In other words, we assume *in-plane* polarization appears in the direction of the *in-plane* director $\mathbf{c}$. Spontaneous polarizations generated from monolayers of these rod-shaped molecules is a vector sum of these electric dipoles. With the orientational distribution function that is an even function of ϕ, $f(\theta, -\phi) = f(\theta, \phi)$, spontaneous polarizations have the form

$$\boldsymbol{P}_0 = \mu N_s[\langle \sin\theta \cos\phi \rangle \mathbf{c}(R) + \langle \cos\theta \rangle \mathbf{n}], \qquad (6.30)$$

where N_s is the area density of constituent molecules, $\langle\ \rangle$ stands for the thermodynamic average with respect to the orientational distribution function, $f(\theta, \phi)$, and $\mathbf{c}(\mathbf{R})$ is *in-plane* director.

We here treat the case of domains with non-uniform distribution of *in-plane* directors accompanied by polarizations, but the magnitudes $P_{0\parallel}(\equiv \mu N_s \sin\theta \cos\phi\rangle)$ of *in-plane* spontaneous polarizations are constants. In this case, Eq. (6.24) is rewritten in the form

$$F_\parallel^{1st} = \frac{1}{2}\int dS k_{ele}(\boldsymbol{R})(\nabla\cdot\boldsymbol{c})^2, \qquad (6.31)$$

with $k_{ele}(\mathbf{R}) = LK(\mathbf{R})P_{0\parallel}^2$.

Note that the contributions from the second term of Eq. (6.24) is neglected because we are treating the case of small gradient of electric charge density in the interior of domains, $s_{0f} = 0$ (see Section 6.2.3).

The second term $F_\parallel$ of Eq. (6.27) has the form

$$F_\parallel^{2nd} = \oint ds a_1(s)\cos\psi(s), \qquad (6.32)$$

with $\cos\psi(s) = \mathbf{c}(\mathbf{r}(s))\cdot\mathbf{m}(s)$ and $a_1(s) = g(s)P_{0\parallel}$. The third term of $F_\parallel$ Eq. (6.29) has the form

$$F_\parallel^{3rd} = \oint \lambda_\parallel(s) + \oint ds a_2(s)\cos 2\psi(s), \qquad (6.33)$$

with

$$\lambda_\parallel(s) = \frac{1}{4}P_{0\parallel}^2\log\frac{L}{\Delta} + \frac{1}{192}\kappa^2 L^2 P_{0\parallel}^2 - \frac{1}{16}P_{0\parallel}^2 L^2\psi_s^2$$

and

$$a_2(s) = \frac{1}{4}P_{0\parallel}^2\log\frac{L}{\Delta} - \frac{1}{192}\kappa^2 L^2 P_{0\parallel}^2 + \frac{1}{16}P_{0\parallel}^2 L^2\psi_s^2,$$

where ψ_s is the first derivative of ψ with respect to the contour length s.

Substituting Eqs. (6.31)–(6.33) into (6.21) leads to the free energy of domains in monolayers in the form

$$F = \Delta P\int dS + \frac{1}{2}\int dS k_{ele}(R)(\boldsymbol{\nabla}\cdot\mathbf{c})^2 + \oint ds\lambda_{ele}(s). \qquad (6.34)$$

Here $\lambda_{ele}(s)$ is anisotropic line tension that has the form

$$\lambda_{ele}(s) = a_0(s) + a_1(s)\cos\psi + a_2(s)\cos 2\psi, \qquad (6.35)$$

with $a_0(s) = \lambda_0 - \frac{\mu_\perp^2}{2}\log(L/\Delta) + \alpha_\perp\kappa^2 + \lambda_\parallel(s)$ and the explicit forms of $a_1(s)$ and $a_2(s)$ which are shown below Eqs. (6.32) and (6.33).

k_{ele} corresponds to Frank elastic constant with respect to splay deformations and its explicit form of this equation is shown below Eq. (6.31), by using *in-plane* polarizations, $P_{0\|}$.

Equation (6.34) is analogous to the form of free energy used by Rudnick and Bruinsma, see also Eq. (6.1). We here note that $(\nabla\Theta)^2$ in Eq. (6.1) is $(\nabla\cdot\mathbf{c})^2+(\nabla\times\mathbf{c})^2$, corresponding to the case of domains where Frank elastic constants for splay and bend deformations are equal. This implies that electrostatic energy arising from *in-plane* dipole–dipole interactions are acting as main origins of the physical mechanisms involved in the 2D Frank elastic energy and anisotropic line tensions of domains in monolayers (see Chapter 5).

In the free energy used by Rudnick and Bruinsma [7], Eq. (6.1), anisotropic line tensions are expanded in the form of Fourier series, see Eq. (6.2) with respect to ψ (that is the angle between in-plane directors $\mathbf{c}$ and normal $\mathbf{m}(s)$), and approximately neglected higher order terms than the second-order $\cos 2\psi$. Anisotropic line tensions, Eq. (6.35), that we derived here have the isotropic term and the anisotropic terms that are proportional to $\cos\psi$ and $\cos 2\psi$, but higher order terms than $\cos 2\psi$ are not allowed. This is because of the principle of superposition of electrostatics, where electrostatic energy of a group of electric charges is the sum of electrostatic energy arising from two-body Coulomb interactions (and thus electrostatic energy has up to the terms $\sim(\mathbf{P}_{0\|}(\mathbf{r}(s))\cdot\mathbf{m}(s))^2$. The fact that the virtual boojum and the cusps of circular domains in fatty acid monolayers are predicted only by anisotropic line tensions up to the term of $\cos 2\psi$ may be because electrostatic energy arising from in-plane dipole–dipole interactions is a key physical mechanism involved in the anisotropic line tensions of domains in monolayers.

Equation (6.1) has both 2D Frank elastic energy contributions for splay and bend deformations, whereas Eq. (6.34) has only Frank elastic energy arising from splay deformations. This is because of the facts that *in-plane* directors are parallel to *in-plane* spontaneous polarizations, and that induced charges are generated by the splay deformations of *in-plane* spontaneous polarizations, but not by their bend deformations. Indeed, Rivière and Meunier showed that asymmetry in Frank elastic constants for splay and bend deformations predicts the cusps of circular domains in fatty acid monolayers [8] and they obtained the best fit between theories and experiments for the case that Frank elastic constant for bend deformations is zero (though the errors of their experiments are relatively large and these authors estimated the values of $(k_s - k_b)/(k_s + k_b)$ can be in the range of 0.3 and 1). This may be because electrostatic energy arising from *in-plane* spontaneous polarizations contributes largely to the 2D Frank elastic energy of domains in fatty acid monolayers.

By the way, nematic liquid crystals in bulk do not show non-centrosymmetric orientational structures, i.e., $S_1 = 0$, and their Frank elastic energies do not have the contributions of electrostatic energy arising from spontaneous polarizations. The Frank elastic energy of bulk nematic liquid crystals is typically in the order of $\sim 10^{-11}$ J. Following the treatments by Helfrich [16, 17], the Frank elastic constants of monolayers are estimated to be $\sim 10^{-20}$ J by multiplying the thickness of monolayers, ~ 1 nm, to the values of Frank elastic constants of bulk nematic liquid crystals, where these Frank elastic constants include the contributions other than electrostatic energy arising from spontaneous polarizations. The contributions of electrostatic energy arising from *in-plane* dipole–dipole interactions are estimated to be $k_{ele} = KLP_{0\parallel}^2 \sim 10^{-19}$ J for the case of $L \sim 50 \sim \mu$m, $P_{0\parallel} \sim 5 \times 10^{-13} \sim$ C/m, and $K \sim 1$; this estimation implies that the contributions of electrostatic energy arising from *in-plane* dipole–dipole interactions relatively large in comparison with other contributions (e.g., anisotropic van der Waals interactions and anisotropic hard-core repulsive interactions). This implies that electrostatic energy arising from *in-plane* dipole–dipole interactions contribute to stabilize the cusps of circular domains in fatty acid monolayers. However, it should be noted that we here used an approximate treatment in the way as in Chapter 5.

6.3. Electrostatic Maxwell Stress Model [18]

6.3.1. *Electrostatic Maxwell stress in domains*

Electrostatic energy arising from *in-plane* dipole–dipole interactions bring 2D Frank elastic energy and anisotropic line tensions. Free energy, approximated through expansions with respect to curvatures, appears to underestimate electrostatic energy resulting from dipole–dipole interactions. Moreover, this free energy presents a contradiction with the laws of electrostatics. Torque balance equations for *in-plane* spontaneous polarizations, derived without curvature expansions, are automatically satisfied at the domain boundary when these in-plane spontaneous polarizations fulfill the torque balance equation in the bulk. However, this does not apply to torque balance equations derived from approximated free energy. Nevertheless, performing calculations for the double area integrals in electrostatic energy arising from *in-plane* dipole–dipole interactions, as given in Eq. (6.21), without curvature expansions becomes challenging, especially when in-plane spontaneous polarizations are distributed non-uniformly.

To reduce this mathematical difficulty, we here revisit free energies of domains in monolayers and drive the general equation that shapes of domains satisfies, by applying the variational principle to the free energy of domains in monolayers that has the form

$$F = \Delta P \int dS + \lambda_0 \oint ds + F_\perp + F_\parallel. \qquad (6.36)$$

Here $F_\perp$ is electrostatic energy arising from normal dipole–dipole interactions (that has the form of the second term of Eq. (4.13) and $F_\parallel$ is electrostatic energy arising from *in-plane* dipole–dipole interactions (that has the form of Eq. (4.46)).

Before applying the variational principle to Eq. (6.36), we present a simple physical argument that leads to the form satisfying the general relation concerning domain shapes. For the case of 2D materials that do not show non-centrosymmetric orientational structures, $S_1 = 0$, electrostatic energy resulting from dipole–dipole interactions does not contribute to the shapes of these materials, $F_\perp = 0$ and $F_\parallel = 0$. In this case, the shapes of these materials are determined by 2D Laplace equation that has the form

$$\Delta P - \lambda_0 \kappa = 0. \qquad (6.37)$$

The solution of Eq. (6.37) gives a circle of radius $r_0 = -\lambda_0/\Delta P$. Laplace equation, indeed, is the balance of pressure difference ΔP and line tensions λ_0 at the boundaries of materials. For the case of domains in monolayers exhibiting non-centrosymmetric orientational structures, $S_1 \neq 0$, electrostatic energy arising from dipole–dipole interactions can contribute to the shapes of these domains as electrostatic *Maxwell stresses*. Maxwell stresses are derived from the principle of virtual displacement which states that stresses applied to

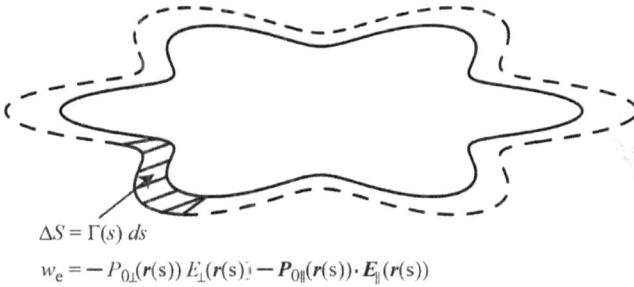

$$\Delta S = \Gamma(s)\, ds$$

$$w_e = -P_{0\perp}(r(s))\, E_\perp(r(s)) - P_{0\parallel}(r(s)) \cdot E_\parallel(r(s))$$

Fig. 6.2: The principle of virtual displacement states that stresses applied to a line segment of a boundary are the decrease of free energy arising from a small displacement of the boundary. A small displacement $\Gamma(s)$ of the boundary of a domain increases electrostatic energy by $\oint w_e \Gamma(s) ds$, where w_e is the area density of electrostatic energy that has the form of Eq. (6.38).

a boundary are the decrease of the energy of the system by a small displacement of this boundary (see Fig. 6.2).

The calculation process is lengthy, but it is detailed in Appendix E (see Appendix E). With a small displacement $\Gamma(s)$ of the boundary in the direction of the boundary normal $\mathbf{m}(s)$, electrostatic energy from dipole–dipole interactions increases by $\oint ds\Gamma(s)w_e$, where $ds\Gamma(s)$ $(= \Delta S)$ is the element of areas that increase by the small displacement $\Gamma(s)$, and w_e is the area density of electrostatic energy that has the form

$$w_e = -\mathbf{P_{0\perp}} \cdot \mathbf{E}_\perp(\mathbf{r}(s)) - \mathbf{P}_{0\parallel}(\mathbf{r}(s)) \cdot \mathbf{E}_\parallel(\mathbf{r}(s)), \qquad (6.38)$$

where $\mathbf{E}_\perp(\mathbf{r}(s))$ and $\mathbf{E}_\parallel(\mathbf{r}(s))$ are the normal and *in-plane* components of (internal) electric fields at $\mathbf{r}(s)$ and have the forms

$$\mathbf{E}_\perp(\mathbf{R}_i) = -\mathbf{P}_{0\perp} \int ds_j \frac{1}{|\mathbf{R}_i - \mathbf{R}_j|^3} \qquad (6.39)$$

$$\mathbf{E}_\parallel(\mathbf{R}_i) = \int ds_j \frac{\mathbf{R}_i - \mathbf{R}_j}{|\mathbf{R}_i - \mathbf{R}_j|^3}(-\mathbf{\nabla}_j \cdot \mathbf{P}_{0\parallel}(\mathbf{R}_j)),$$

$$+ \oint ds_j \frac{\mathbf{R}_i - \mathbf{r}(s_j)}{|\mathbf{R}_i - \mathbf{r}(s_j)|^3} \mathbf{P}_{0\parallel}(\mathbf{r}(s_j)) \cdot \mathbf{m}(s_j). \qquad (6.40)$$

Using Maxwell stresses that have the form of Eq. (6.38), the general relation that is satisfied by the general shape of domains is derived in the form [18]

$$\Delta P - \lambda_0 \kappa - P_{0\perp} E_\perp(\mathbf{r}(s)) - \mathbf{P}_{0\parallel}(\mathbf{r}(s)) \cdot \mathbf{E}_\parallel(\mathbf{r}(s)) = 0. \qquad (6.41)$$

It is important to note that Maxwell stresses, the third and fourth terms of Eq. (6.41) have the forms that are different from the forms, $\frac{1}{2}\epsilon E^2$. This is because Maxwell stresses are driven by normal and *in-plane spontaneous* polarizations (and thus have the form of the area density of energy stored by dipole–dipole interactions). In simpler terms, the electrostatic dipolar energy stemming from dipole–dipole interactions acts as the origin of Maxwell stress within monolayer domains. Consequently, the equilibrium between line tensions and electrostatic Maxwell stress determines the shapes of these domains. Given that *in-plane* directors are distributed non-uniformly in a domain, it implies a non-uniform distribution of *in-plane* spontaneous polarizations. These non-uniform *in-plane* spontaneous polarizations are automatically included in Eq. (6.41) with electrostatic Maxwell stress. Equation (6.41) indicates that Maxwell stress serves as the source of the additionally negative line tension and a curvature-elastic energy, as derived in the geometric shape equation in Chapter 5. This conclusion is pivotal as it underlies the physics of the analytical method elucidated in Chapter 5.

6.3.2.   *Domain shape analyses using electrostatic Maxwell stress*

Certainly, it would be beneficial to illustrate how Eq. (6.41) can be employed to analyze domain shapes, despite not being expressed in terms of geometric parameters. Prior to delving into the analysis of domain shapes in monolayers while considering Maxwell stress, it is useful to reframe Eq. (6.41) in the following form:

$$\Delta P - \lambda_0 \kappa - \mathbf{M}_{0\perp} \cdot \mathbf{B}(\mathbf{r}(s)) - \mathbf{P}_{0\parallel}(\mathbf{r}(s)) \cdot \mathbf{E}_\parallel(\mathbf{r}(s)) = 0. \qquad (6.42)$$

with $\mathbf{M}_{0\perp} = \mathbf{M}_0 \mathbf{n}$,
and

$$\mathbf{B}(\mathbf{R}) = \oint ds_j \frac{P_{0\perp} \mathbf{t}(s_j) \times (\mathbf{R}_i - \mathbf{r}(s_j))}{|\mathbf{R}_i - \mathbf{r}(s_j)|^3}. \qquad (6.43)$$

The first, second, and fourth terms of Eq. (6.42) are identical to the first, second, and fourth terms of Eq. (6.41). The third term

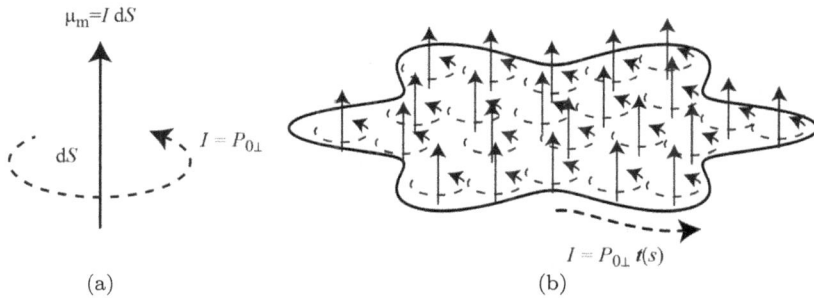

$\mu_m = I\,dS$

dS

$I = P_{0\perp}$

$I = P_{0\perp}\,t(s)$

(a)   (b)

Fig. 6.3:   Analogy with 2D magnetic domains. The general shape equation of domains in monolayers, Eq. (6.41), is rewritten in the form of Eq. (6.42), where the third term has the form of Maxwell stresses arising from magneto static interactions between spontaneous magnetizations $\mathbf{M}_0 = P_{0\perp}\mathbf{n}$ and magnetic flux density $\mathbf{B}(\mathbf{r}(s))$ that has the form of Eq. (6.42).

has the form of Maxwell stresses arising from magnetic dipoles: In 2D magnetic domains of uniform (spontaneous) magnetizations $\mathbf{M}_0 (= P_{0\perp}\mathbf{n})$, loop electric currents that generate these magnetizations are canceled out at the internal parts of the domains and thus only electric currents along the boundary of the domain generate magnetic flux density, see Fig. 6.3. Magnetic flux density arising from electric currents, $\mathbf{I}(\equiv P_{0\perp}\mathbf{t}(s))$, flowing along the boundary has the form of Eq. (6.43) (Biot–Savart's law). Magneto-static energy generated by magneto-static interactions between magnetic dipoles $\mathbf{M}_0 dS$ at an area element and magnetic flux density at this position has the form of $-(\mathbf{M}_0 dS) \cdot \mathbf{B}(\mathbf{R})$. Applying the principle of virtual displacement, see also Fig. 6.2, leads to the third term of (6.42). The physical meaning of Eq. (6.41) is clear, but it is mathematically convenient to use Eq. (6.42) because only a single line integral is involved.

6.4.   Analyses of Cusps in Circular Domains

In this section, an example of domain shape analysis using Eqs. (6.41) and (6.42) is provided for the case of cusps. The liquid crystal theory of monolayers suggests that spontaneous splay induces virtual boojum *in-plane* director distributions within domains of fatty acid monolayers, and anisotropic line tension stabilizes cusps in their circular domains [7]. Analyzing these cusps in domains of fatty acid monolayers using Eq. (6.41), which includes terms related to electrostatic Maxwell stress, is of interest. Since the shapes of domains in fatty acid monolayers are mostly circular, the contribution of Maxwell stresses arising from in-plane spontaneous polarizations acts as perturbations. Assuming an appropriate solution, akin to the approach adopted by McConnell and coworkers [19] in Chapter 5, we can investigate the case of cusps. This involves *a priori* assuming shapes as solutions and subsequently verifying their validity.

Indeed, the most probable solution of Eq. (6.42) can be expressed as $\mathbf{r}(s) = r_0(\cos\theta, \sin\theta) + g(\theta)(\cos\theta, \sin\theta)$, where $g(\theta)$ signifies the impact of Maxwell stresses arising from in-plane spontaneous polarizations or dipole–dipole interactions. In more detail, when Maxwell stresses from in-plane spontaneous polarizations are negligible or minor (no perturbations), a circular domain with a radius of r_0 is expected to be a solution to Eq. (6.42) because the third and fourth terms related to polarizations, leading to Maxwell stresses, are neglected. Therefore, $g(\theta)$ essentially represents the influence of Maxwell stresses caused by polarizations.

In this case, "fictitious" magnetic flux densities, Eq. (6.43), at the boundary of this domain have the form

$$B_\perp^{(0)} = P_{0\perp} r_0^{-1} \log\left(\frac{8r_0}{\Delta e}\right). \tag{6.44}$$

Substituting Eq. (6.44) and the curvature of circular domains into Eq. (6.42) leads to the form

$$\Delta P + \frac{\lambda_0}{r_0} - \frac{P_{0\perp}}{r_0} \log\left(\frac{8r_0}{\Delta e}\right) = 0. \tag{6.45}$$

The left-hand side of Eq. (6.45) is indeed equal to the first derivative of the free energy, Eq. (5.16), with respect to r_0, except for the contributions of difference, ΔP; Eq. (6.45) predicts the optimal radius of circular domains in the same way as the prediction of McConnell–Moy theory (see Chapter 5).

Fictitious magnetic flux densities are calculated by the first-order term with respect to $g(\theta)$ in the form

$$B_\perp^{(1)} = \sum_{n=0}^{\infty} B_n c_n r_0^{-1} \cos n\theta, \tag{6.46}$$

with

$$B_0 = P_{0\perp} r_0^{-1} \log \frac{e^2 \Delta}{8r_0},$$ (6.47)

$$B_n = 2P_{0\perp} r_0^{-1} \left[-(n^2 - 1) \log \frac{e^2 \Delta}{8r_0} - \frac{n^2 - 1}{2} \right.$$

$$\left. + \frac{4n^2 - 1}{4} \left(\psi \left(\frac{3}{2} \right) - \psi \left(n + \frac{1}{2} \right) \right) \right] \; (n \neq 0).$$ (6.48)

We expanded $g(\theta)$ in a Fourier series, see Eq. (4.17), and the phases of the expansion coefficients c_n (that is in general a series of complex numbers), $n = 0, 1, \ldots$, are fixed to $c_n = c_{-n}$; this represents a shape that is symmetric with respect to the x-direction (that is the reference axis to measure θ). Noteworthy that B_n are related to polarizations $P_{0\perp}$.

Following the mathematical treatment of differential geometry, see Section 5.1, leads to the curvatures of Eq. (5.18) in the form

$$\kappa^{(1)} = -\frac{1}{r_0} \sum_{n=-\infty}^{\infty} (n^2 - 1) c_n r_0^{-1} e^{in\theta},$$ (6.49)

where only the first-order term with respect to $g(\theta)$ is shown. Substituting Eqs. (6.46) and (6.48) into Eq. (6.42) leads to the form

$$\delta p + P_{0\perp}^2 r_0^{-1} \sum_{n=-\infty}^{\infty} \Omega_n^\perp c_n r_0^{-1} e^{in\theta} = 0,$$ (6.50)

where δp is first-order pressure differences that ensure the area of the domain does not change with $g(\theta)$. The explicit form of $\Omega_n^\perp$ is shown in Eq. (5.21) in Chapter 5.

The left-hand side of Eq. (6.50) is indeed the first-order derivative of Eq. (5.16) with respect to c_n. $\Omega_n^\perp$ is a function of n and radius r_0; $\Omega_n^\perp$ takes positive values for relatively small radius r_0 and changes to negative values with increasing radius r_0 at a threshold value r_n that is dependent on n, see also Section 5.2.

When all coefficients $\Omega_n^\perp$, $n = 0, 1, 2, \ldots$, are positive, $c_n = 0$ is the only solution of Eq. (6.50); in this case, Maxwell stresses that have tendency to destabilize circular domains are suppressed by line tensions. In contrast, for the cases that one of these coefficients, say $\Omega_n^\perp$, is negative, line tensions cannot suppress Maxwell stresses arising from normal spontaneous polarizations (here, we remind the readers that the subscript n of $\Omega_n^{\perp}$ corresponds to the Fourier mode of the expansion of $g(\theta)$, see Eq. (5.17). In this case, circular domains are no longer stable with respect to the Fourier mode n of shape deformations $g(\theta)$ (where the corresponding coefficient $\Omega_n^{\perp}$ is negative) and thus circular domains show "shape" transitions to clover shaped domains with n lobes; this also agrees with the results of McConnell–Moy theory [10]. Threshold domain radius r_n, at which the coefficients $\Omega_n^{\perp}$ change sign, are increasing function of n, $r_n < r_{n+1}$, and thus the first mode that shows instability is $n = 2$. Because we treat the cases that the shapes of domains are circular when Maxwell stresses arising from in-plane spontaneous polarizations are absent, in this section, we treat the cases of domains, whose radii are smaller than $r_2 \left(= \frac{\Delta}{8} e^{\lambda_0/P_{0\perp}+10/3} \right)$.

To calculate Maxwell stresses arising from *in-plane* spontaneous polarizations, it is necessary to relate *in-plane* spontaneous polarizations and *in-plane* directors. We thus use a molecular model that is used in Section 5.5; we treat fatty acids as cylinders that have electric dipoles at their center of gravity, where these electric dipoles are directed toward the long-axes of cylinders. Fatty acid molecules are achiral and have mirror symmetry with respect to tilt plane (that is made by the direction of tilting and interface normal). Spontaneous polarizations thus have the form of Eq. (6.30); *in-plane* spontaneous polarizations are indeed parallel to *in-plane* directors, $\mathbf{P}_{0\parallel}(\mathbf{R}) = P_{0\parallel}\mathbf{c}(\mathbf{R})$. We here treat the case that the magnitudes of *in-plane* spontaneous polarizations $P_{0\parallel}$ is uniform. In-plane directors show virtual boojum distributions that have the form of Eq. (6.9); this treats the cases that defects that generate *in-plane* director distributions are located at $(x_d, 0)$ that is the exterior of the domains. Electric charges induced in the boundary and interior of a domain thus have the form, see Eqs. (6.22) and (6.23),

$$\sigma_{\text{ind}}(\mathbf{r}(s)) = P_{0\parallel} \frac{(r_0^2 + x_d^2)\cos\theta - 2r_0 x_d}{r_0^2 + x_d^2 - 2r_0 x_d \cos\theta}$$

$$= P_{0\parallel} \left[-\frac{r_0}{x_d} + \left(1 - \left(\frac{r_0}{x_d}\right)^2 \right) \sum_{n=1}^{\infty} \left(\frac{r_0}{x_d}\right)^{n-1} \cos n\theta \right],$$

$$(6.51)$$

$$\rho_{\text{ind}}(\mathbf{R}) = -2P_{0\parallel}\frac{\cos\phi_d}{r_d}$$

$$= \frac{2P_{0\parallel}}{x_d}\left(1 + \sum_{n=1}^{\infty}\left(\frac{r}{x_d}\right)^n \cos n\theta\right). \tag{6.52}$$

The following mathematical formula

$$\frac{1 - b^2}{1 - 2b\cos\theta + b^2} = 1 + 2\sum_{n=1}^{\infty} b^n \cos n, \tag{6.53}$$

was used to derive the last forms of Eqs. (6.51) and (6.52).

Substituting Eqs. (6.51) and (6.52) into Eq. (6.40) leads to *in-plane* electric fields (that are parallel to the interfaces) at domain boundaries in the form

$$E_{\parallel}(r(s)) = P_{0\parallel}r_0^{-1}\left[\frac{r_0}{x_d}(2\alpha_1 - \gamma_1)(\cos\theta, \sin\theta)\right.$$

$$+ \frac{1}{2}\sum_{n=1}^{\infty}\left[\left(1 - \left(\frac{r_0}{x_d}\right)^2\right)\gamma_{n+1} + 2\left(\frac{r_0}{x_d}\right)^2 \alpha_{n+1}\right]$$

$$\times \left(\frac{r_0}{x_d}\right)^{n-1}(\cos(n+1)\theta, \sin(n+1)\theta)$$

$$+ \frac{1}{2}\sum_{n=1}^{\infty}\left[-\left(1 - \left(\frac{r_0}{x_d}\right)^2\right)\gamma_n + 2\left(\frac{r_0}{x_d}\right)^2 \beta_{n-1}\right]$$

$$\left. \times \left(\frac{r_0}{x_d}\right)^{n-1}(\cos(n-1)\theta, -\sin(n-1)\theta)\right], \tag{6.54}$$

by the leading order terms with

$$\alpha_{n+1} = \int_0^1 dt \int_{-\pi}^{\pi} d\theta \frac{t^{n+1}\cos n\theta - t^{n+2}\cos(n+1)\theta}{\left(1 + t^2 - 2t\cos\theta + \frac{\Delta^2}{r_0^2}\right)^{\frac{3}{2}}}. \tag{6.55}$$

$$\beta_{n-1} = \int_0^1 dt \int_{-\pi}^{\pi} d\theta \frac{t^{n+1}\cos n\theta - t^{n+2}\cos(n-1)\theta}{\left(1 + t^2 - 2t\cos\theta + \frac{\Delta^2}{r_0^2}\right)^{\frac{3}{2}}}, \tag{6.56}$$

$$\gamma_{n+1} = \int_{-\pi}^{\pi} d\theta \frac{\cos n\theta - \cos(n+1)\theta}{\left(2 - 2\cos\theta + \frac{\Delta^2}{r_0^2}\right)^{\frac{3}{2}}}. \tag{6.57}$$

Interactions between these in-plane electric fields and in-plane spcn-taneous polarizations generate Maxwell stresses that have the form

$$w_\parallel = -P_{0\parallel} \sum_{n=0}^{\infty} E_n \cos n\theta. \tag{6.58}$$

These Maxwell stresses contribute to domain shape as follows (see Eq. (6.50));

$$\delta p - (\lambda_0 r_0^{-1} + P_{0\perp} B_0) \frac{c_0}{r_0} - P_{0\parallel} E_0$$

$$+ 2P_{0\perp}^2 r_0^{-1} \sum_{n=1}^{\infty} \Omega_n^{\perp} c_n r_0^{-1} \cos n\theta - P_{0\parallel} \sum_{n=1}^{\infty} E_n \cos n\theta = 0, \tag{6.59}$$

where only the first-order terms with respect to $g(\theta)$ and $w_\parallel$ are shown.

The solution of Eq. (6.59) thus has the form

$$\delta p = (\lambda_0 r_0^{-1} + \mathbf{P}_{0\perp} \mathbf{B}_0) r_0^{-1} c_0 + \mathbf{P}_{0\parallel} \mathbf{E}_0, \tag{6.60}$$

$$\frac{c_n}{r_0} = \frac{\mathrm{P}_{0\parallel}^2}{\mathrm{P}_{0\perp}^2} \frac{\tilde{E}_n}{2\Omega_n^{\perp}} \quad (n \geq 2), \tag{6.61}$$

with $\tilde{E}_n = E_n/(P_{0\parallel} r_0^{-1})$.

Numerical calculations of Eq. (6.61) show the presence of *quasi*-cusps, see Fig. 6.4; tangent vectors are continuous along *quasi*-cusps and are discontinuous at (real) cusps. At a first glance, one may think that this result is the artifact of using Eq. (5.17) to represent shapes because trigonal functions $e^{in\theta}$ are continuous functions. However, Eq. (5.17) can indeed represent a discontinuous function when the coefficients c_n are independent of n;

$$\sum_{n=-\infty}^{\infty} e^{in\theta} = 2\pi \delta(\theta). \tag{6.62}$$

The Fourier coefficients E_n of Maxwell stresses (and thus the Fourier components c_n of the shapes of domains) decrease exponentially with increasing n (see Eq. (6.61)).

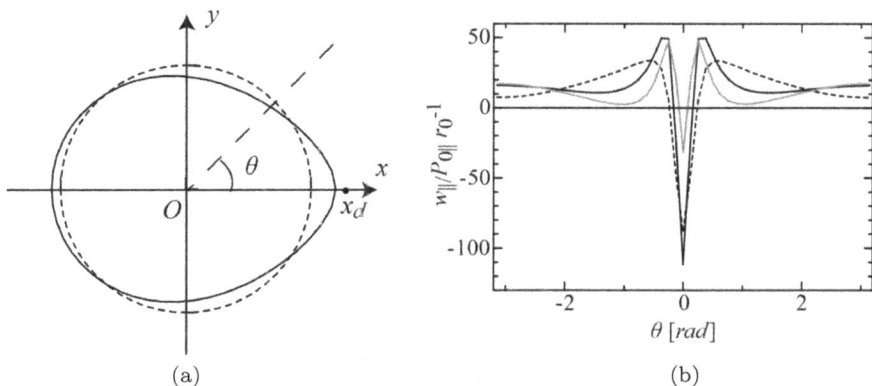

(a)   (b)

Fig. 6.4: Equation (6.61) is numerically calculated for $\lambda_0/P_{0\perp}^2 = 9.19$, $P_{0\parallel}^2/P_{0\perp}^2 = 0.02$, $\Delta/r_0 = 3.0 \times 10^{-5}$, and $r_0/x_d = 0.78$ (a). These values correspond to, for example, $r_0 = 40 \sim \mu$m, $\Delta = 1.2 \sim$ nm, $P_{0\perp} = 2.6 \times 10^{-12}$ C/m, $P_{0\parallel} = 3.7 \times 10^{-13}$ C/m, and $\lambda_0 = 5.6 \times 10^{-13}$ N. For these values of parameters, Maxwell stresses arising from induced electric charges in the interior (the black broken curve) and boundary (the gray solid curve) of the domains and the sum of them (the black solid curve) are calculated (b). [Reproduced from Ref. [18].]

In summary, by considering the contribution of electrostatic Maxwell stress, it becomes evident that domains with *in-plane* directors exhibiting virtual boojum distributions only manifest quasi-cusps instead of real cusps. Experimentally, whether domains in fatty acid monolayers stabilize (real) cusps or quasi-cusps is a controversial issue: This suggests that the fine details of these structures are not conclusively determined because the observation of the fine structures of domains is limited by the resolution of microscopes (see Fig. 6.1).

Positive electric charges are induced in the interior of the domain, concentrating near the vicinity of the x-axis (refer to Fig. 6.5(a)). These induced positive electric charges generate in-plane electric fields directed toward the exterior of the domain. These in-plane electric fields apply Maxwell stresses to the positively charged parts of the boundary near the virtual defect (as shown in Fig. 6.5(b)), stabilizing the quasi-cusp of domains in fatty acid monolayers. The Maxwell stress model provides insights into the physical mechanisms that stabilize the shapes of domains through in-plane electric fields.

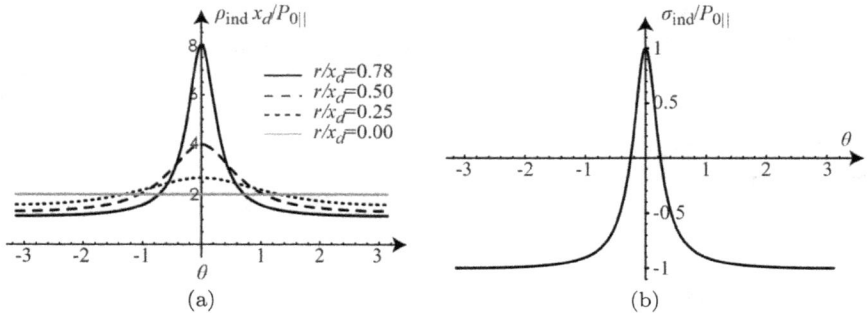

(a)   (b)

Fig. 6.5: The densities of induced electric charges at the interior (a) and boundary (b) of domains are calculated for $\lambda_0/P_{0\perp}^2 = 9.19$, $P_{0\|}^2/P_{0\perp}^2 = 0.02$, $\Delta/r_0 = 3.0 \times 10^{-5}$, and $r_0/x_d = 0.78$ (the same values as those used in Fig. 6.4 (4.8)).

Domains exhibit quasi-cusps for specific values of a/x_d, and these values are determined by minimizing the free energy with respect to a/x_d. The free energy is expanded by the second-order term with respect to c_n and $P_{0\|}^2$;

$$F = F_0 + F_{\mathrm{ori}} + F_{\mathrm{sha}}, \qquad (6.63)$$

where F_0 is the free energy contributions due to line tensions and electrostatic energy arising from normal spontaneous polarizations for circular domains. F_{ori} is the free energy contributions due to the virtual boojum distributions of in-plane spontaneous polarizations and F_{sha} is the free energy contributions due to the deformations of the shapes of the domain. F_{ori} and F_{sha} has the form

$$\frac{F_{\mathrm{ori}}}{\pi r_0^2 (P_{0\|}^2 r_0^{-1})}$$

$$= 4\left[\left(\frac{r_0}{x_d}\right)^2 \int_0^1 dt_i \int_0^1 dt_j t_i t_j L_0(t_i, t_j) \right.$$

$$\left. + \frac{1}{2} \sum_{n=1}^{\infty} \left(\frac{r_0}{x_d}\right)^{2n+2} \int_0^1 dt_i \int_0^1 dt_j t_i^{n+1} t_j^{n+1} L_n(t_i, t_j) \right]$$

$$+\left[\left(\frac{r_0}{x_d}\right)^2 L_0(1,1) + \frac{1}{2}\sum_{n=1}^{\infty} L_n(1,1)\left(\left(\frac{r_0}{x_d}\right)^{n-1} - \left(\frac{r_0}{x_d}\right)^{n+1}\right)\right]$$

$$+4\left[-\left(\frac{r_0}{x_d}\right)^2\left(\int_0^1 dtt L_0(t,1)\right)\right.$$

$$\left.+\frac{1}{2}\sum_{n=1}^{\infty}\left(\frac{r_0}{x_d}\right)^{2n}\left(1-\left(\frac{r_0}{x_d}\right)^2\right)\left(\int dtt^{n+1}L_n(t,1)\right)\right], \qquad (6.64)$$

$$\frac{F_{\text{sha}}}{\pi r_0^2(P_{0\|}^2 r_0^{-1})} = -\left[\frac{\lambda_0}{P_{0\perp}^2} - \log\left(\frac{8r_0}{e^2\Delta}\right)\right]\frac{P_{0\|}^2}{P_{0\perp}^2}\sum_{n=2}^{\infty}\left[\frac{\tilde{E}_n}{2\Omega_n^\perp}\right]^2$$

$$-\frac{1}{2}\frac{P_{0|}^2}{P_{0-}^2}\sum_{n=2}^{\infty}\frac{\tilde{E}_n^2}{\Omega_n} + \tilde{E}_0. \qquad (6.65)$$

For the cases that the values of $\lambda_0/P_{0\perp}^2$ are not too large, the free energy is minimum at $r_0/x_d \sim 0.78$, where we predicted that domains stabilize *quasi*-cusp (see Fig. 6.6). The general equation provided by Eq. (6.41) predicts that domains in fatty acid monolayers stabilize quasi-cusps, which qualitatively agrees with experimental

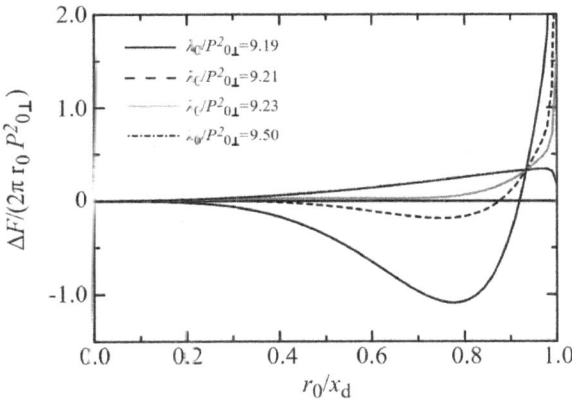

Fig. 6.6: The free energy is shown as a function of r_0/x_d for $\lambda_0/P_{0\perp}^2 = 9.19$ (black solid curve), 9.21 (black broken curve), 9.23 (gray solid curve), 9.50 (black chain curve). The values of $P_{\text{c}|}^2/P_{0\perp}^2$ and Δ/r_0 are fixed to 0.02 and 3.0×10^{-5}, respectively.

observations. This offers one potential mechanism through which circular domains in fatty acid monolayers stabilize quasi-cusps.

6.5.   Summary

Molecules in domains of condensed phase show local orientational order with respect to their tilting directions (local *in-plane* orientational order) and their average orientations are represented by *in-plane* directors, and domains of condensed phase in monolayers show non-uniform *in-plane* director distributions. This situation is analogous to liquid crystals, which are easily deformed by small perturbations due to boundary conditions, applied fields, etc. Domains of condensed phase that coexist with domains of liquid expanded phase in fatty acid monolayers show non-uniform *in-plane* director distributions, called virtual boojum distributions. Domains in fatty acid monolayers comprised of rod-like achiral molecules show circular shapes, where cusps are stabilized at the vicinity of virtual defects that generate virtual boojum in-plane director distributions. Employing liquid crystal theories is effective to account for the stabilization of the cusps of domains in fatty acid monolayers, but insufficient to reach solid conclusions regarding the stabilization.

Domains of condensed phase show local in-plane orientational order and their in-plane directors are distributed non-uniformly. This implies that in-plane spontaneous polarizations (that are parallel to interfaces) are generated in these domains due to the peculiar polar property of the constituent amphiphilic molecules in monolayers, and their orientations are non-uniform. Consequently, electrostatic energy is stored in the domains. Electrostatic energy arising from in-plane spontaneous polarizations is rewritten in the form of electrostatic energy arising from electrostatic Coulomb interactions between polarization charges (or bound charges) induced at the interior $\rho_{\text{ind}} = -\nabla \cdot \mathbf{P}_0(\mathbf{R})$ and the boundary $\sigma_{\text{ind}} \equiv \mathbf{P}_0(\mathbf{r}(s)) \cdot \mathbf{m}(s)$ of the domains.

By using expansions based on Frenet–Serret theorem, the free energy of domains in monolayers is written in an analogous form to the free energy of the 2D Frank elastic energy of nematic liquid crystals by Rudnick and Bruinsma. However, this approach is still insufficient and cannot quantitatively predict reasonable results.

Going back to the concept of shape equations, the general relation concerning domain shapes are derived by applying the variational principle to the free energy of domains in monolayers. Electrostatic Maxwell stresses is shown to play an important role in stabilizing domain shapes, and gives the balance of forces (per unit length) applied to the boundary of domains; electrostatic energy arising from spontaneous polarizations contributes to the shape of domains. This model can account for the presence of *quasi*-cusps of domains in fatty acids, where virtual boojum distributions induce positive polarization charges in the interior of the domain and generate electric fields toward the exterior of the domain. These electric fields apply forces to the positively charged parts of the domain boundary (that is close to the virtual defect) toward the exterior of the domain; these forces stabilize the *quasi*-cusp. Nevertheless, it should be critically tested in future experiments, with consideration of the elastic liquid crystalline energy of domains with in-plane ordering, besides the electrostatic Maxwell stress energy (see Chapter 7).

In essence, a comprehensive approach for analyzing domain shapes, considering both the elastic Frank energy and electrostatic dipolar interactions simultaneously, will be necessary. This involves modeling a molecule with two distinct parts: polar heads related to polar properties and long chains related to liquid crystalline properties. This allows for the separate treatment of the principal sources of polar and liquid crystalline properties. The analytical method using this model is described in Chapter 7, where shape equations using geometric parameters are treated in the manner as outlined in Chapter 5, and this method is well utilized to show the presence of variety domains with intriguing shapes.

References

[1]   R. M. Weis and H. M. McConnell, Two-dimensional chiral crystals of phospholipid, *Nature*, 310, 47–49 (1984).

[2]   V. T. Moy, D. J. Keller, and H. M. McConnell, Molecular order in finite two-dimensional crystals of lipid at the air–water interface, *J. Phys. Chem.*, 92, 5233–5238 (1988).

[3]   V. T. Moy, D. J. Keller, H. E. Gaub, and H. M. McConnell, Long-range molecular orientational order in monolayer solid domains of phospho-lipid, *J. Phys. Chem.*, 90, 3198–3202 (1986).

[4]   P. Krüger and M. Lösche, Molecular chirality and domain shapes in lipid monolayers on aqueous surfaces, *Phys. Rev. E*, 62, 7031 (2000).

[5]   D. K. Schwartz, M. W. Tsao, and C. M. Knobler, Domain morphology in a two-dimensional anisotropic mesophase: Cusps and boojum textures in a Langmuir monolayer, *J. Chem. Phys.*, 101, 8258–8261 (1994).

[6]   T. M. Fischer, R. F. Bruinsma, and C. M. Knobler, Textures of surfactant monolayers, *Phys. Rev. E*, 50, 413–428 (1994).

[7]   J. Rudnick and R. Bruinsma, Shape of domains in two-dimensional systems: Virtual singularities and a generalized Wulff construction, *Phys. Rev. Letts.*, 74, 2491–2494 (1995).

[8]   S. Rivière and J. Meunier, Anisotropy of the line tension and bulk elasticity in two-dimensional drops of a mesophase, *Phys. Rev. Letts.*, 74, 2495–2498 (1995).

[9]   D. Petty and T. C. Lubensky, Stability of texture and shape of circular domains of Langmuir monolayers, *Phys. Rev. E*, 59, 1834–1845 (1999).

[10] H. M. McConnell and V. T. Moy, Shapes of finite two-dimensional lipid domains, *J. Phys. Chem.*, 92, 4520–4525 (1988).

[11] M. Iwamoto and Z. C. Ou-Yang, Shape deformation and circle instability in two-dimensional lipid domains by dipolar force: A shape- and size-dependent line tension model, *Phys. Rev. Lett.*, 93(20), 206101 (2004).

[12] M. Iwamoto, F. Liu, and Z. C. Ou-Yang, Shape and stability of two-dimensional lipid domains with dipole–dipole interactions, *J. Chem. Phys.*, 125, 224701 (2006).

[13] T. Yamamoto, T. Manaka, and M. Iwamoto, The interacting electric charge model on the shape formation of monolayer domains at the air–water interface comprised of tilted dipoles with orientational deformation, *Thin Solid Films*, 516, 2660–2665 (2008).

[14] T. Yamamoto, T. Manaka, and M. Iwamoto, Electrostatic origin of the Frank elastic energy and anisotropic line tension of the domains in mono-layers at the air–water interface, *Euro. Phys. J. E*, 29, 1–8 (2009).

[15] T. Yamamoto, *Dielectric and Geometrical Properties of Dipolar Monolayers in Liquid Crystal Phase*, Tokyo Institute of Technology, Tokyo, September 2007.

[16] Z. C. Ou-Yang and W. Helfrich, Instability and deformation of a spherical vesicle by pressure, *Phys. Rev. Lett.*, 59, 2486 (1987).

[17] Z. C. Ou-Yang, J. X. Liu, and Y. Z. Xie, *Geometric Methods in the Elastic Theory of Membranes in Liquid Crystal Phases*, World Scientific, Singapore, 1999.

[18]   T. Yamamoto, D. Taguchi, M. Weis, T. Manaka, and M. Iwamoto, Electrostatic Maxwell stress model of the shapes of condensed phase domains in monolayers at the air-water interface, *J. Chem. Phys.*, 128, 204706 1–21 (2008).

[19]   H. M. McConnell and V. T. Moy, Shapes of finite two-dimensional lipid domains, *J. Phys. Chem.*, 92, 4520–4525 (1988).

Chapter 7

Comprehensive Approach for Analyzing Domain Shapes

This chapter outlines a comprehensive methodology for analyzing domain shapes in lipid monolayers, focusing on distinct molecular structures contributing to polar and liquid crystalline properties. The approach separates the treatment of polar heads, associated with polar properties, and long chains, linked to liquid crystalline properties. The modeling of domain shapes in lipid monolayers is based on the concept of polar cholesteric liquid crystals [1]. Analytical solutions are presented, with consideration of both the elastic Frank energy and electrostatic dipolar interactions. The study introduces an evident and analytic shape solution for pinned boundary orientation that considers maximum boundary tension. This solution successfully explains characteristic domain shapes observed in lipid monolayers, including star, boojum, cardioid, ellipse, bola, and clover-leaf shapes (refer to Section 7.1).

Moreover, the chapter delves into the texture transformation within circular domains of polar smectic films. This analysis incorporates chiral elasticity resulting from the coupling of flexoelectric and spontaneous polarizations (refer to Section 7.2). The consideration of these factors provides insights into the observed transformations and contributes to a more comprehensive understanding of the behavior of lipid monolayers in different conditions.

7.1.   Shapes of Tilted Lipid Domains

7.1.1.   *Comprehensive approach for analyzing tilted lipid domains*

As described in Chapter 5, according to the development of many sophisticated microscopy techniques, such as fluorescence microscopy (FM) [2, 3] and Brewster angle microscopy (BAM) [4], a variety of domain shapes from round structures to complex labyrinthine patterns [5–11] have been found and many studies have been conducted to understand the formation mechanisms of such domain shapes. There are at least two independent approaches for interpreting domain features. The first one is to understand the domain shapes from the peculiar polar property of monolayers of amphiphilic molecules at the air–water interface, where solid domains (LC phase) surrounded by fluid phase (LE phase) observed by FM and BAM is treated as an equilibrium state due to the competition between line tension and electrostatic repulsion due to dipole–dipole interaction [7] (see Chapter 5). In previous chapters, the mathematical approach using *differential geometry* is introduced for analyzing such domain shapes (see Chapter 5), from which the double line integral of the dipolar energy due to dipole–dipole interaction is interpreted as the sum of additionally negative line tension and curvature-elastic energy of the domain boundary. The second one is to understand domains from their inner textures, on focusing the crystalline nature of domains, and the molecular orientation in domains detected by BAM is analyzed based on the two-dimensional (2D) Frank elasticity theory of liquid crystals, e.g., chiral smectic liquid crystals [CSLC] [12]. In Chapter 6, it is shown that the double line integral of the dipolar energy is the source of the Maxwell stress, and the case of monolayer domains possessing the liquid-crystalline nature is treated, where *in-plane* spontaneous polarization is postulated to be formed along the director fields. Using the *in-plane* polarization model based on this idea, *quasi*-cusps of domains in fatty acids were analyzed, with consideration of the Maxwell stress. However, this approach is merely one way to account for domain shapes. It will be necessary to pay more attention to actual molecular structures and properties and develop a method for analyzing domain shapes, in which the effect of both polar property and liquid-crystalline property is included because there are a variety of domain shapes that cannot be treated by employing approaches described in Chapters 5 and 6. In this chapter, we describe the comprehensive approach, where a molecular model is modified from the models that are used in Chapters 5 and 6 by paying more attention to actual molecular structures.

A formalism is conducted by taking account of dipole–dipole interaction of the terminal methyl groups and carbonyls on the molecular double hydrocarbon chains and their orientation order. Both bulk and boundary orientations and boundary-shape equations for tilted lipid domain are derived in analogy with a polar cholesteric liquid crystal (PCLC). Here the "polar" has diploid meanings in physics, the dipole–dipole interaction is incorporated into LC Frank free energy and the D_∞ symmetry of CSLC has been broken into C_∞ for PCLC, i.e., the Frank free energy has no invariance of $d \to -d$ [13], where d is the director, which is described by the orientation of the hydrocarbon chains.

It shows that in 2D system, the domain, the 3D spontaneous splay and chiral elastic energies, the s_0 and k_2 terms of Frank energy, can be regarded as an *orientation-dependent line tension* at the boundary and the domain formation is the equilibrium between the line tension, the surface pressure, and the orientational stress, and the dipole–dipole interaction. It will be shown that there is an obvious and analytic shape solution for pinned boundary orientation for maximum boundary tension, and the solution describes well the diverse domain shapes, such as star, boojum, cardioid, ellipse, bola, and clover-leaf shapes [5–11]. In other words, the comprehensive approach described here can reveal a variety of domain shapes in monolayers.

7.1.2. *Modeling of tilted lipid domains and basic equations for analyzing shapes*

The domain is modeled with a close boundary $\mathbf{r}(x, y) = \mathbf{r}(s)$, and a 3D director $d = (\sin\theta_0 \cos\phi(x, y), \sin\theta_0 \sin\phi(x, y), \cos\theta_0)$ to describe the internal tilted orientation, where θ_0 is the constant tilted angle of d to the normal z of the water surface (see Fig. 7.1). The internal tilted orientation represents the liquid crystal property of the molecule.

On the other hand, the "polar" cholesteric liquid crystal nature is also treated by the "polar" head of the molecule. As a result, the

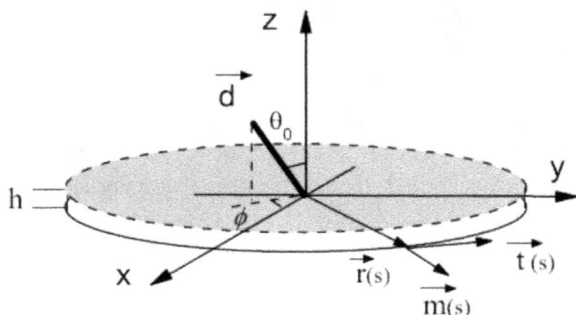

Fig. 7.1: A schematic diagram of the shape of a monolayer domain and the geometric quantities. The thick line represents the hydrocarbon chains in the domain.

Source: Reprinted from Fig. 1 in Ref. [1], *Europhys. Lett.*, 91, 16004, 2010.

free energy of domains is described by a sum of four terms as

$$F = \Delta P \int dA + \gamma \oint ds - \frac{\mu^2}{2} \oint \oint \frac{\boldsymbol{t}(l) \cdot \boldsymbol{t}(s)}{|\boldsymbol{r}(l) - \boldsymbol{r}(s)|} dl ds + F_{\mathrm{PCLC}}. \quad (7.1)$$

The first three terms have been used to discuss the domain formation with the approach used in previous chapters [7, 14, and Chapter 5]: $\Delta P = \Pi - g_0$, Π is the surface pressure and g_0 the difference in the Gibbs free energy density between the outer (isotropic fluid) and inner (now, PCLC); γ the line tension; the third term represents the dipole–dipole interaction between the inherent dipoles of the "polar" head of lipid molecules [15], μ is the dipole density, $\boldsymbol{t} = d\boldsymbol{r}/ds = (\cos \Phi(s), \sin \Phi(s))$ is the unit tangential vector of the boundary and $\Phi(s)$ is the boundary orientational angle at the arc length s. Here the dipole–dipole interaction is considered only for the dipolar component normal to the water surface, because the dipolar moments in chains and headgroups behave entirely different, in the manner as seen in the analysis in Ref. [16]. Both dipolar moments give rise to image dipoles: terminal methyl groups and carbonyls on the chains are embedded in a medium of $\epsilon \approx 2$, the normal components of their image dipoles have the *same* directions as of the real dipole moments. Therefore, the effective dipole moments from both real and image dipoles are *normal* to the interface, giving rise to the interaction shown in the third term of Eq. (7.1). By contrast, head group dipole moments are located in water with dielectric permittivity $\epsilon \approx 78 (\gg 10)$ and normal projection of their image dipoles have *opposite* direction to the real ones, but resulting in uncanceled normal effective dipole moments. The last term in Eq. (7.1), FPCLC, is the elastic Frank energy [13]

$$F_{\text{PCLC}} = \frac{1}{2} \int [k_{11}(\nabla' \cdot \boldsymbol{d} - s_0)^2 + k_{22}(\boldsymbol{d} \cdot \nabla' \times \boldsymbol{d} - k_2)$$
$$+ k_{33}(\boldsymbol{d} \times \nabla' \times \boldsymbol{d})^2] \mathrm{d}V, \tag{7.2}$$

where $\nabla' = (\partial_x, \partial_y, \partial_z)$. $\boldsymbol{d}$ describes the internal tilted orientation of molecules, where the long-chains of molecules make the main contribution.

With a straight calculation and neglecting a constant which can be incorporated in ΔP, Eq. (7.2) can be expressed in the Langer–Sethna form [12] as

$$F_{\text{PCLC}} = \frac{1}{2} \int [K_s(\nabla \cdot \boldsymbol{c})^2 + K_b(\nabla \times \boldsymbol{c})^2] \mathrm{d}A$$
$$+ \oint [\lambda_s(\boldsymbol{c} \cdot \boldsymbol{m}) + \lambda_b(\boldsymbol{c} \times \boldsymbol{m})] \mathrm{d}l, \tag{7.3}$$

where $\boldsymbol{c} = (\cos \phi(x, y), \sin \phi(x, y))$ is the 2D director, $\nabla = (\partial_x, \partial_y)$, $K_s = k_{11}h\sin^2\theta_0/2$, $K_b = h[k_{11}^2 \cos\theta_0 + k_{33}^2 \sin\theta_0]\sin^2\theta_0/2$, $\lambda_s = -s_0 k_{11}h\sin\theta_0$, $\lambda_b = k_2 k_{22}h\sin\theta_0\cos\theta_0$, $\boldsymbol{m} = (\sin\Phi(s), -\cos\Phi(s))$, is the outer ward normal to the boundary, and h is the thickness of the monolayer.

The advantage of present derivation from 3D Frank energy, the above relationships, gives the estimates of the 2D splay and bending curvature elastic constants K_s, $K_b \approx 10^{-12}$ erg with the same magnitude as the membrane bending rigidity derived by Helfrich [17]. Especially, it shows that the bulk splay and chiral elastic energy associated with s_0 and k_2 are the sources of the anisotropic line tensions of λ_s and λ_b.

For reducing the mathematical complex without loss of the generality of physics here, one-constant approximation, i.e., $k_{11} = k_{22} = k_{33} = k$, is employed. Then Eq. (7.3) becomes

$$F_{\text{PCLC}} = \int \frac{K}{2} |\nabla \cdot \phi|^2 \, \mathrm{d}A + + \oint \sigma(\Phi - \phi)\mathrm{d}l. \tag{7.4}$$

Here, $K = kh\sin^2\theta_0$, $\sigma(\Phi - \phi) = -\overline{\lambda}\cos(\Phi - \phi - \alpha)$
with $\overline{\lambda} = \sqrt{\lambda_s^2 + \lambda_b^2}$, $\sin\alpha = -\lambda_s/\sqrt{\lambda_s^2 + \lambda_b^2}$,
and $\cos\alpha = \lambda_b/\sqrt{\lambda_s^2 + \lambda_b^2}$.

To find the optimal domain shape and texture, F is minimized with respect to $\phi(x, y)$ and $\boldsymbol{r}(s)$.

By parameterizing $\phi'(x, y) = \phi(x, y) + \delta\phi(x, y)$ and $\boldsymbol{r}' = \boldsymbol{r}(s) + \psi(s)\boldsymbol{m}(s)$ where $\delta\phi$ and ψ are infinitesimal various functions. From Eqs. (7.1) and (7.4), the variation of F is derived in the form as

$$\delta F = F' - F = \int [\cdots]_1 \delta\phi dA + \oint [\cdots]_2 \delta\phi dl + \oint [\cdots]_3 \psi dl + O(\delta\phi)^2 +$$
$O(\delta\phi\psi)$, with $[\cdots]_1 = 0$, $[\cdots]_2 = 0$, and $[\cdots]_3 = 0$.

Then the bulk orientation, boundary orientation, and boundary shape equation are obtained, respectively, as

$$\Delta\phi = 0, \tag{7.5}$$

$$K\vec{m}\cdot\nabla\phi - \overline{\lambda}\sin(\Phi - \phi - \alpha) = 0, \tag{7.6}$$

$$\frac{K}{2}|\nabla\phi|^2 - \overline{\lambda}\cos(\Phi - \phi - \alpha)\boldsymbol{t}\cdot\nabla\phi - \overline{\lambda}\sin(\Phi - \phi - \alpha)\boldsymbol{m}$$

$$\cdot\nabla\phi + \Delta P - \lambda\kappa + \alpha_0\kappa^3 + 2\alpha_0\kappa_{ss} = 0, \tag{7.7}$$

where $\overline{\lambda} = \gamma - (\mu^2/2)\ln(L/eh) + (11/48)L\mu^2 \oint \kappa^2 ds$, $\alpha_0 = (11/96)$ $\mu^2 L^2$, $\kappa = -d\Phi/ds$ is the curvature of the boundary, $\kappa_{ss} = d^2\kappa/ds^2$, and L is the boundary length [14, 15].

Equations (7.5)–(7.7) are represented using geometric parameters, and they provide basic equations for analyzing domain shapes.

7.1.3. *Shape equations for analyzing domains*

We need to solve Eqs. (7.5)–(7.7) to determine domain shapes, but it is an intractable task to find the general solution for $\phi(x, y)$ and $\boldsymbol{r}(s)$, owing to highly nonlinear property of Eqs. (7.6) and (7.7). It is therefore better to find some special solutions by considering relevant physics on monolayer domains. As we know from $\overline{\lambda}$ expression, the electrostatic line forces tend to expand the domain size and boundary, and for forming a stable domain a positive line tension is crucial.

Therefore, a maximum σ, the orientation-induced line tension, looks like a condition for a compact domain.

It is found from Eq. (7.4) $\Phi - \phi - \alpha = \pi$ makes

$$\sigma = \sigma_{\max} = \sqrt{\lambda_s^2 + \lambda_b^2}. \tag{7.8}$$

Moreover, the pinned boundary condition (7.8) predicts a correct orientation. Using definitions of α, λ_s, λ_b, we perform a check:

$-\pi/2 < \alpha < 0$ for s_0, $k_2 > 0$ and c is left-handed outer-ward to the boundary;

$-\pi < \alpha < -\pi/2$ for $s_0 > 0$, $k_2 < 0$ and c is right-handed outer-ward;

$0 < \alpha < \pi/2$ for $s_0 < 0$, $k_2 > 0$ and c is left-handed inward;

$\pi/2 < \alpha < \pi$ for s_0, $k_2 < 0$ and c is right-handed inward.

All show good agreement with optimal orientation to minimize Frank energy, i.e., $k_2 > 0$ prefers c left-handed orientation and $k_2 < 0$ makes c right-handed one as seen from Eq. (7.2). The condition given by Eq. (7.8) reveals two limiting cases: strong chirality pinned c parallel to the boundary as proposed in Ref. [12] and confirmed by experiment in smectic-C* [18], while strong spontaneous splay ($\lambda_s \gg \lambda_b$) tends to orient molecules normal to the boundary.

Substituting Eq. (7.8) into Eq. (7.6) yields $m \cdot \nabla\phi = 0$, $|\nabla\phi|^2 = |t \cdot \nabla\phi|^2$ and $t \cdot \nabla\phi = \cos\Phi \partial\phi/\partial x + \sin\Phi \partial\phi/\partial y = \mathrm{d}\Phi/\mathrm{d}s = -\kappa(s)$. The relations then reduce Eq. (7.7) into the following concise form

$$\Delta P - (\overline{\lambda} + \lambda)\kappa + \frac{K}{2}\kappa^2 + \alpha_0\kappa^3 + 2\alpha_0\kappa_{ss} = 0. \tag{7.9}$$

Equation (7.9) has similar forms which are derived in Chapter 5, and this can be a starting point to analyze a variety of shapes of domains, because Eq. (7.9) is the equation comprising geometric parameters κ and κ_{ss}.

7.1.4.   *Solutions of shape equations*

Using elliptic functions one can exactly integrate Eq. (7.9), in a manner as it has been done for case of $K = 0$ [19] (see Chapter 5). For a given solution $r(s)$ or $\Phi(s)$, one can have the $\phi(x,y)$ value at the boundary from Eq. (7.8), and with the so-called first boundary value

$\phi(s)$ (Eq. (7.5)) then gives the unique inner harmonic function $\phi(x, y)$ (the famous Dirichlet problem). In one word, the pinned boundary condition (7.8) is a close form in mathematics.

An obvious conclusion from Eq. (7.9) is that in pure *smectic-C** (i.e., $\mu = 0$ and $\alpha_0 = 0$) the domain shapes must be circular. This is again confirmed by Meyer's group [18]. As a consequence, Eq. (7.9) predicts that non-circular shapes cannot be explained by line tension anisotropy alone. Of course, the conclusion is not the case of $K_s \neq K_b$ which was invoked to simulate the pentadecanoic acid boojum shape by anisotropic line tension alone [20] (see Chapter 6).

In what follows, using Eq. (7.9) we can prove non-circular shapes without assuming $K_s \neq K_b$. It is obvious that a circle of radius ρ_0 is always a solution of Eq. (7.9) if its curvature $\kappa_0 = -1/\rho_0$ satisfies the following equation:

$$\Delta P - (\overline{\lambda} + \lambda)\kappa_0 + \frac{K}{2}\kappa_0^2 + \alpha_0\kappa_0^3 = 0. \tag{7.10}$$

From the equation we can evaluate ρ_0 from both elastic property $(K, \overline{\lambda})$ and polar one (α_0, λ). To search for a non-circular solution, the mentioned integration using elliptic functions is only formalism and it is difficult to show a clear feature. Therefore, we turn to consider a slightly distorted circle with curvature of $\kappa = \kappa_0 + \delta\kappa |\delta\kappa| \ll |\kappa_0|$. Substituting to Eq. (7.9) and only taking the first order of δk, we obtain the following differential equation:

$$\frac{d^2}{ds^2}\delta\kappa + q^2\delta\kappa = 0, \tag{7.11}$$

where $q^2 = [K\kappa_0 - (\overline{\lambda} + \lambda) + 3\alpha_0\kappa_0^2]/2\alpha_0$.

From the above equation, we have $\delta\kappa = a\cos qs + b\sin qs$, if $q^2 > 0$, i.e., $\rho_0 \geq \left[\sqrt{K^2 + 12\alpha_0(\overline{\lambda} + \lambda)} - K\right]/[2(\overline{\lambda} + \lambda)]$.

For any close curves, there are, at least, four extreme points at which $d\kappa/ds = 0$ (the four-vertex theorem). By choosing $s = 0$ at an extreme point, we have $\delta\kappa = a\cos qs$. Considering $\delta k(0) = \delta k(L)$, with $L \approx 2\pi\rho_0$ we have $q = n/\rho_0$ ($n = 1, 2, \ldots$). From the relation (7.10), and the relation of q with K, κ_0, we can derive the critical pressure ΔP_n at which the nth harmonic shape happens.

ΔP_n will be a complex function of material parameters K, s_0, k_2, and μ. From $-d\Phi/ds = \kappa_0 + \delta$ we finally have the nth harmonic shape solution

$$\Phi'(s) = \frac{s}{\rho_0} + \frac{a\rho_0}{n} \sin \frac{ns}{\rho_0}, \qquad (7.12)$$

and the domain shape $r(s) = [x(s), y(s)]$ as

$$x(s) = \int_0^s \cos \Phi(s')ds',$$

$$y(s) = \int_0^s \sin \Phi(s')ds', \qquad (7.13)$$

where the parameter a can be determined by a given $\kappa(0)$ (see Eq. (7.9)). To compare with experimental observation topologically, in what follows, we set $\rho_0 = 1$ for numerical simulation of Eq. (7.13).

Figure 7.2 shows the results of $n = 1$, $a = 0.2$. Such boojum and cardioid shapes were indeed observed simultaneously at the $\Pi - A$ plateau of LE-LC cc-existence in the diamitoyl-phosphadidylcholine (DPPC) experiment (Fig. 1b in Ref. [6]). Both shapes become the common forms observed in DPPC monolayers [2–11] in the beginning of distortion of circle by compression.

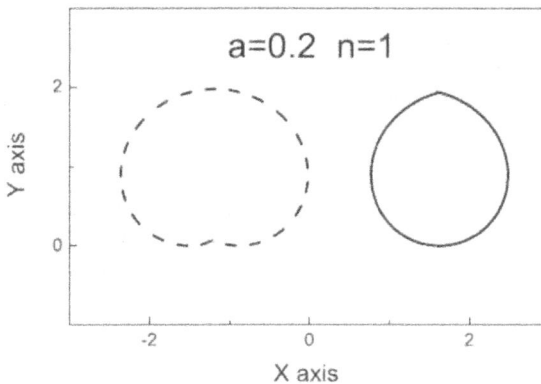

Fig. 7.2: Calculation of domain shapes by the solution of Eqs. (7.12) and (7.13) with $n = 1$, $\rho_0 = 1$ and $a = 0.2$. Both they are commonly observed at the beginning at the surface pressure plateau of LE–LC coexistence and has been called boojum (right) and cardioid (left) domains.

Source: Reprinted from Fig. 2 in Ref. [1], *Europhys. Lett.*, 91, 16004, 2010.

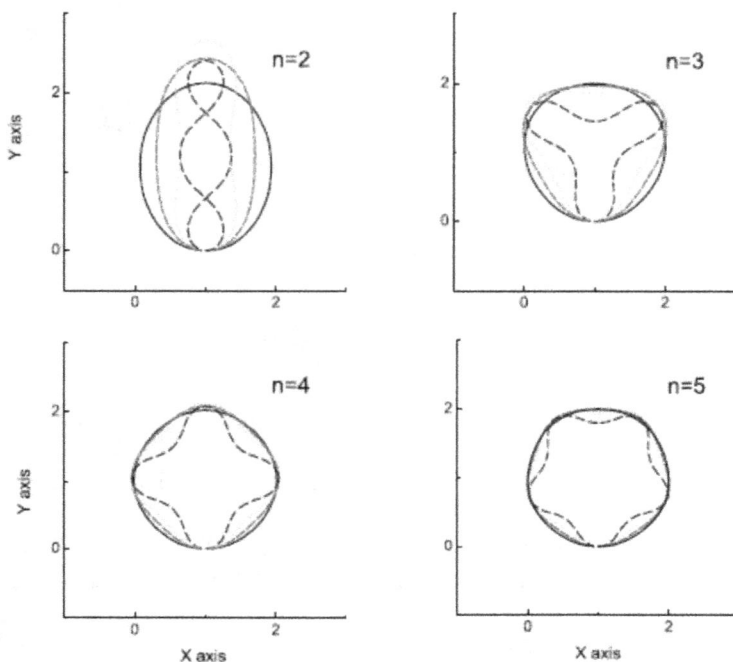

Fig. 7.3:   Illustration of circle undulation of mode $n = 2$; 3; 4; 5 calculated by Eqs. (7.12) and (7.13) with $\rho_0 = 1$, $a = 0.2$ (solid line), 0.8 (dash line), 1.6 (dot line), and 3.2 (dot-dash line). A close inspection shows a symmetry break with regular undulation, e.g., Fig. 2 in Ref. [10]. The latter has exact Cn symmetry but present figure shows a distorted Cn symmetry but keeps $C2$ symmetry. The symmetry break is in good agreement with the experiments, e.g., Fig. 1 in Ref. [9]. *Source*: Reprinted from Fig. 3 in Ref. [1], *Europhys. Lett.*, 91, 16004, 2010.

Figure 7.3 shows more results of Eq. (7.12) with $a = 0.2$ (solid line), 0.8 (dash line), 1.6 (dot line), and 3.2 (dot-dash line) for different n-values (2; 3; 4; 5). The figure of $n = 2$ shows a shape transition from circle to ellipse, bola that is again in agreement with experiment; the ellipse appears in Fig. 5 in Ref. [7]. The bola shape is often found in the shear-induced equilibrium domains [5, 21]. The shear-induced circle-to-ellipse deformation has been investigated in theory where the Frank elastic effect has not been invoked [22]. If we combine the effect, we believe the bola deformation must be described as present. Generally speaking, Eq. (7.12) reveals n-mode undulating deformation of circle domain. However, a close inspection shows a

symmetry break of the present result, Fig. 7.3, with regular undulation, e.g., Fig. 2 in Ref. [10]. For the example of $n = 3$, the clover-
leaf-shaped (CLS) domains shown in Fig. 7.3 is not of exact $C3$
symmetry but a distorted triangle with $C2$ symmetry. The agreement of the prediction with experimental observation, such as the
CLS domains given Fig. 1 in Ref. [9] and Fig. 1(c,d) in Ref. [6], is
quite good.

The discussion in Section 7.1 shows that the common and intriguing shapes of lipid monolayer domains can be understood as equilibrium between surface pressure, PCLC Frank elasticity of molecular
hydrocarbon chains, and the dipole–dipole interaction of the molecular polar groups on chains. This may have bio-medicine significance
such as investigation of domain shape to give insight into the interaction of local anesthetics with lipids in membranes [23].

7.2. Texture Transformation in Circular Domain of Polar Smectic Films

In this section, the stability of both pure tangential and radial textures in circular domains of polar SmC* LCs is shortly discussed in
terms of elastic anisotropy. At a critical domain diameter, both textures switch to reversing spiral textures (see Fig. 7.4); by considering
the chiral elasticity contributed by the coupling of flexoelectric and
spontaneous polarizations, the existence of an inverse relation of the
critical radius and the spontaneous polarization can be revealed [24],
by modeling the domain and its inner texture as in Section 7.1.1 [8].

In more detail, theoretical investigations of domain formation
in systems of tilted molecules on air/liquid interfaces have been
actively conducted. The lipid domains are studied as equilibrium
between competing surface pressure, line tension, and electrostatic
repulsion of amphiphilic molecules [7] while the study of Smectic C*
liquid crystal (SmC* LC) domain [12] focuses on the molecular orientation and inner textures based on two-dimensional (2D) Frank
elasticity theory [13]. Although theoretical progress by combining
texture, line tension, and domain shape has been made [20, 25, 26],
electrostatic dipole interactions have not been incorporated in the
theory until the dipole–dipole interaction was shown to be a shape-
and size-dependent line tension [14] (see Chapter 5). However, the

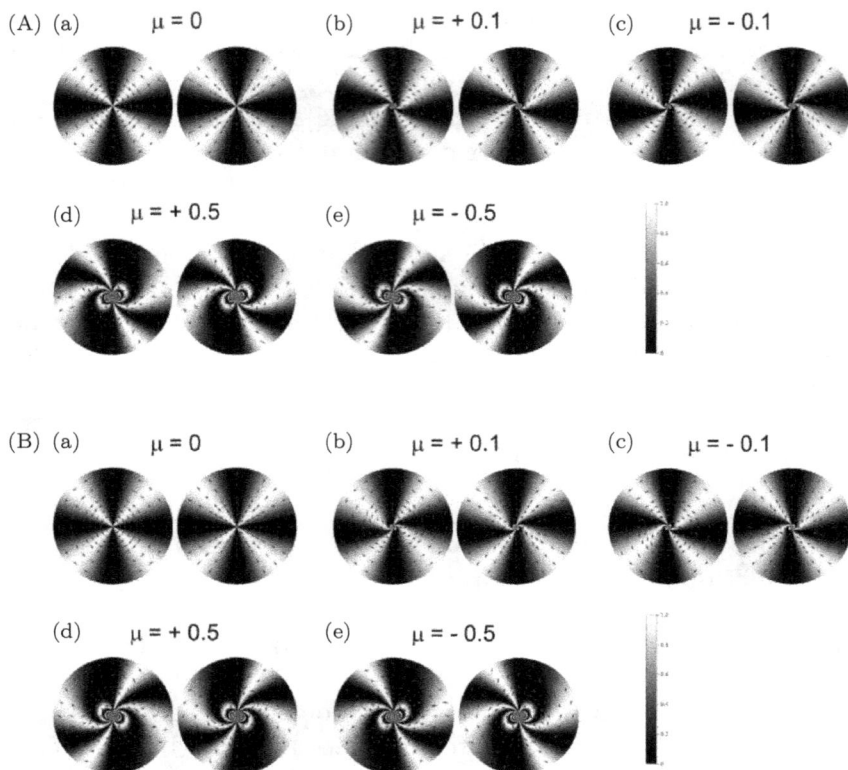

Fig. 7.4: Textures and sketches of director patterns calculated by $I/I_{max} = \sin^2 2\emptyset$ and $\hat{c} = (\cos\phi, \sin\phi)$, respectively: The left sides of (a–e) for 7.4(A), $\phi = \theta + \pi/2 + 2\mu\cos\theta/r$ (for Fig. 7.4(B), $\phi = \theta + \pi/2 - 2\mu\cos\theta$), and right sides for $\phi = \theta + 2\mu\cos\theta/r$ (for 1B, $\phi = \theta - 2\mu\cos\theta/r$) where $\mu = 0, 0.1, -0.1, 0.5, -0.5$, for (a), (b), (c), (d), (e).

Source: Reprinted from Fig. 1 in Ref. [24], *Chem. Phys. Lett.*, 628, 96–100, 2015.

study of domain formation in lipid monolayers [7, 14] is based on the assumption of a homogeneous molecule-tilted or normal domain (see Chapter 5). In Section 7.1 [1], both bulk and boundary orientation, and boundary shape equations for tilted lipid domain have been derived in analogy with a polar cholesteric LC (PCLC) under the equal elastic constant (EECP) approximation. The simplicity of EECP reduces mathematical challenges in the problem, but has the possibility to lose important physical features. For example, the discussion of textural transformation [27] in circular domain of polar smectic films has to invoke elastic anisotropy (see below, for details) [18]. In free-standing films of one SmC* LC, i.e., both polar and chiral PCLC, Meyer *et al.* discovered a beautiful texture transition in circular domains [27]: Small domains tend to have the target-like simple pure tangential texture and, above a certain critical domain diameter, the texture can switch to a reversing spiral. The latter texture was first observed in a domain at an air–water interface [28]. The mechanical process involving forces that generate mechanical stress, strain, and movement of 2D domains and their inner textures has received a great deal of attention in the literature [29, 30]. However, discussions on the texture transformation [12, 13, 18, 20, 25, 26] still rely on frame-type with a variety of assumptions such as EECP [26] and somewhat arbitrary definitions of line tension or strong anchoring at edge [18, 20]. Why there is a critical domain diameter which depends on domain thickness. typically in the range of tens of microns? And why there is an obvious inverse relation between the critical diameter and spontaneous polarization found in a mixture of CS1024 and CS1015 (see Fig. 2 and Table I in Ref. [31])?

Clear answers to these questions are provided in Ref. [24]; modeling the domain and its inner texture as in Section 7.1.1 [1], it is shown that a circular domain associated with simple pure tangential (radial) texture always minimizes the free energy if its normal (tangential) component of line tension is absent. The texture transition from both radial and tangential textures is analytically shown by solving the dynamical equations for the first-order perturbations of the elastic anisotropy (see Fig. 7.4). It is found that at a critical domain diameter, both textures can switch to reversing spiral textures as has been observed [27, 31]. The critical size is obtained exactly as the ratio of a 2D elastic constant and line tension, both of which are functions of film thickness, molecular tilt angle, and 3D elastic constants of the LC. Evaluating this expression approximately shows that the theoretical critical diameter is in the same range as provided by the experiment [27]. By considering the chiral elasticity contributed mainly by the coupling of flexoelectric and spontaneous polarizations, the existence of an inverse relation of the critical radius and the spontaneous polarization can be shown as observed in Ref. [31] (refer to Fig. 2), where the elastic anisotropy is an important parameter for understanding the texture transformations in free-standing SmC* LC films [24]. The expressions obtained for the critical domain diameter may have significance in determining the elastic constants and tilt angles for SmC* materials. The good agreement with previous experimental observations confirms the validity and shows its potential for applications to new hot topics in the investigation of defects in LCs and other soft materials.

References

[1]   M. Iwamoto, F. Liu, and Z.-C. Ou-Yang, Domain shapes in lipid monolayers studied as polar cholesteric liquid crystals, *Europhys. Lett.*, 91, 16004 (2010).

[2]   V. von Tscharner and H. M. McConnell, An alternative view of phospholipid phase behavior at the air–water interface: Microscope and film balance studies, *Biophys. J.*, 36(2), 409–419 (1981).

[3]   M. Lösche, E. Sackmann, and H. Möhwald, A fluorescence microscopic study concerning the phase diagram of phospholipids, *Ber Bunsenges Phys. Chem.*, 87, 848–852 (1983).

[4]   S. Henon and J. Meunier, Microscope at the Brewster angle: Direct observation of first-order phase transitions in monolayers, *Rev. Sci. Instrum.*, 62, 936 (1991).

[5]   D. J. Benvegnu and H. M. McConnel, Line tension between liquid domains in lipid monolayers, *J. Phys. Chem.*, 96, 6820 (1992).

[6]   K. Kjaer, J. Als-Nielsen, C. A. Helm, L. A. Laxhuber, and H. Möhwald, Ordering in lipid monolayers studied by synchrotron X-ray diffraction and fluorescence microscopy, *Phys. Rev. Lett.*, 58, 2224 (1987).

[7]   D. J. Keller, J. P. Korb, and H. M. McConnell, Theory of shape transitions in two-dimensional phospholipid domains, *J. Phys. Chem.*, 91, 6417 (1987).

[8]   H. E. Gaub, V. T. Moy, and H. M. McConnell, Reversible formation of plastic two-dimensional lipid crystals, *J. Phys. Chem.*, 90, 1721 (1986).

[9]   V. T. Moy, D. J. Kelly, and H. M. McConnel, Molecular order in finite two-dimensional crystals of lipid at the air–water interface, *J. Phys. Chem.*, 92, 5233 (1988).

[10]   T. K. Vanderlick and H. Moehwald, Mode selection and shape transitions of phospholipid monolayer domains, *J. Phys. Chem.*, 94, 886 (1990).

[11]   M. Weis and H. M. McConnel, Cholesterol stabilizes the crystal–liquid interface in phospholipid monolayers, *J. Phys. Chem.*, 89, 4453 (1985).

[12] S. A. Langer and J. P. Sethna, Textures in a chiral smectic liquid-crystal film, *Phys. Rev. A*, 34, 5035–5046 (1986).

[13] F. C. Frank, 1. Liquid crystals. On the theory of liquid crystals, *Discuss. Faraday Soc.*, 25, 19 (1958).

[14] M. Iwamoto and Z.-C. Ou-Yang, Shape deformation and circle instability in two-dimensional lipid domains by dipolar force: A shape- and size-dependent line tension model, *Phys. Rev. Lett.*, 93(20), 206101 (2004).

[15] M. Iwamoto, F. Liu, and Z.-C. Ou-Yang, Shape and stability of two-dimensional lipid domains with dipole–dipole interactions, *J. Chem. Phys.*, 125, 224701 (2006).

[16] P. Krüger and M. Lösche, Molecular chirality and domain shapes in lipid monolayers on aqueous surfaces, *Phys. Rev. E*, 62, 7031 (2000).

[17] W. Helfrich, Elastic properties of lipid bilayers: Theory and possible experiments, *Z. Naturforsch. C.*, 28, 693–703 (1973).

[18] J. B. Lee, D. Konovalov, and R. B. Meyer, Textural transformations in islands on free standing smectic-C* liquid crystal films, *Phys. Rev. E*, 73, 051705 (2006).

[19] M. Iwamoto, F. Liu, and Z.-C. Ou-Yang, Shapes of lipid monolayer domains: Solutions using elliptic functions, *Eur. Phys. J. E*, 27, 81–86 (2008).

[20] K. K. Loh and J. Rundnick, Textures and the shapes of domains in Langmuir monolayers, *Phys. Rev. Lett.*, 81, 4935–4938 (1998).

[21] S. Trabelsi, S. Zhang, T. Randall Lee, and D. K. Schwartz, Linactants: Surfactant analogues in two dimensions, *Phys. Rev. Lett.*, 100, 037802 (2008).

[22] M. Iwamoto, T. Yamamoto, F. Liu, and Z.-C. Ou-Yang, Shear-induced domain deformation in a tilted lipid monolayer: From circle to ellipse and kinked stripe, *Phys. Rev. E*, 78, 051704 (2008).

[23] S. A. Kane and S. D. Floyd, Interaction of local anesthetics with phospholipids in Langmuir monolayers, *Phys. Rev. E*, 62, 8400 (2000).

[24] M. Iwamoto, D. Taguchi. and Z.-C. Ou-Yang, Texture transformation in circular domain of polar smectic films: Chiral elasticity induced by coupling of flexoelectric and spontaneous polarizations, *Chem. Phys. Lett.*, 628, 96–100 (2015).

[25] J. Rundnick and R. Bruinsma, Shape of domains in two-dimensional systems: Virtual singularities and a generalized Wulff construction, *Phys. Rev. Lett.*, 74, 2491 (1995).

[26] D. Pettey, T. C. Lubensky, and D. R. Link, Topological inclusions in 2D smectic C films, *Liq. Cryst.*, 25, 579–589 (1998).

[27] R. B. Meyer, D. Konovalov, I. Kraus, and J. B. Lee, Equilibrium size and textures of islands in free-standing smectic C* films, *Mol. Cryst. Liq. Cryst. Sci. Tech. Sec. A* 364, 123–131 (2001).

[28] I. Kraus and R. B. Meyer, Polar smectic films, *Phys. Rev. Lett.*, 82, 3815 (1999).

[29] C. Bohley and R. Stannarius, Inclusions in free standing smectic liquid crystal films, *Soft Matter*, 4, 683 (2008).

[30] Y.-K. Kim, S. V. Shiyanovski, and O. Laventovich, Morphogenesis of defects and tactoids during isotropic–nematic phase transition in self-assembled lyotropic chromonic liquid crystals, *J. Phys. Condens. Matter*, 25, 404202 (2013).

[31] J.-B. Lee, R. A. Pelcovits, and R. B. Meyer, Role of electrostatics in the texture of islands in free-standing ferroelectric liquid crystal films, *Phys. Rev. E*, 75, 051701 (2007).

Appendix A

Electrostatic Energy and Dielectric Polarizations of Dielectric Materials

Using orientational order parameters S_n $(n = 0, 1, 2, \ldots)$, isotropic material systems are characterized as $S_1 = 0$, $S_2 = 0$, and $S_3 = 0$, meanwhile monolayer systems are characterized as $S_1 \neq 0$, $S_2 \neq 0$, and $S_3 \neq 0$. The knowledge of isotropic materials is helpful in treating monolayers. In this appendix polarizations and electrostatic energy of isotropic dielectric materials are summarized based on the classical electromagnetism theory.

A.1. Dielectric Polarizations of Isotropic Bulk Materials and Capacitance

Ideal isotropic dielectric materials do not allow conductive currents to flow [1,2] but can induce electric charges. Induced charges are *bound charges*, and called *polarization charges* because they are induced due to dielectric polarizations of materials (Purcell proposed to call them "structural charges" [3]). Induced charges cannot be extracted outside. For the case of uniform dielectric materials, induced charges appear only at their surfaces, see Fig. A.1(a), whereas for the case of non-uniform dielectric materials, induced charges also appear in their bulk (see Fig. A.1(b)).

Parallel plate capacitors are useful to outline the physics of dielectric polarizations, see Fig. A.1(c), where dielectric material is sandwiched between parallel planar electrodes. Under applied voltages V,

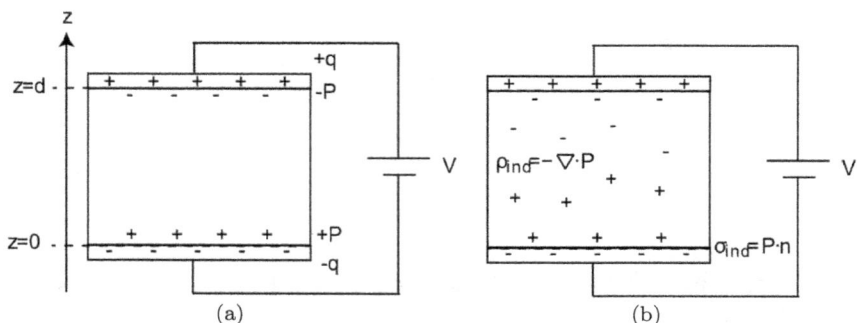

Fig. A.1:   Dielectric polarizations are generated in isotropic bulk materials when external electric fields are applied. Consequently, electric charges are induced on the surface of the materials, but not in their bulk (a). For non-uniform dielectric materials, electric charges are induced not only on the surface of these materials but also in their bulk (b).

electric charges are induced on the electrodes, $+Q$ and $-Q$, in proportion to V. The coefficient of the proportionality defined by [4, 5]

$$C = \frac{Q}{V}, \tag{A.1}$$

characterizes the quantity of capacitors (capacitance). Capacitance C depends on the intrinsic dielectric properties of dielectric materials sandwiched between the parallel electrodes. In applied electric fields (external electric fields), *bound charges* appear uniform only on the surface of isotropic dielectric material, see Fig. A.1(a), meanwhile dielectric polarizations are formed in the dielectric material in proportion to external electric fields $\boldsymbol{E}$ as

$$\boldsymbol{P} = \epsilon_0 \chi^{(1)} \boldsymbol{E}. \tag{A.2}$$

Here the dimensionless coefficient $\chi^{(1)}$ is called electric susceptibility and it is constant. $\epsilon_0 (= 8.85 \times 10^{-12}$ [C^2/m^{2}N or F/m]) is dielectric constant of vacuum. Dielectric polarizations $\boldsymbol{P}$[C/m^2] defined by Eq. (A.2) are called linear (dielectric) polarizations. Electric fields are uniform in the dielectric material, and they are given by

$$E = -(q - P)/\varepsilon_0, \tag{A.3}$$

where q is the area density of electric charges induced on the electrodes, see Fig. A.1. The capacitance of parallel plate capacitors has the form

$$C = \frac{qA}{V} = \epsilon \frac{A}{d}, \tag{A.4}$$

with

$$\epsilon = \epsilon_0 (1 + \chi^{(1)}), \tag{A.5}$$

where A and d, respectively, are the areas of the two electrodes and the distance between them. ϵ is called dielectric constant. The dielectric properties of isotropic materials are characterized using dielectric constant ϵ (or susceptibility $\chi^{(1)}$).

A.2. Electrostatic Energy [4, 5]

Dielectric polarizations increase the electrostatic energy of the system. The energy dW that is necessary to convey electric charges $d\tilde{Q}$ from one electrode to the counter electrode via the external circuit, has the form $dW = V d\tilde{Q}$, where V is the difference of potentials between the two electrodes. Now we consider a process of gradually transporting electric charges on the electrode of this capacitor from $\tilde{Q} = 0$ to $\tilde{Q} = Q$ (the electric charges on the counter electrode change from $\tilde{Q} = 0$ to $\tilde{Q} = -Q$); this process leads to the electrostatic energy W in the form

$$W = \int_0^Q V d\tilde{Q} = Q^2/2C. \tag{A.6}$$

Equation (A.6) is rewritten in the form

$$W = \frac{1}{2}\epsilon E^2 A d, \tag{A.7}$$

using Eqs. (A.4) with $E = Vd$ and $Q = qA$. The volume density of electrostatic energy has the form $w_e = \frac{1}{2}\epsilon E^2$, see Eq. (A.7), where Ad is the volume of dielectric materials. Using Eq. (A.5), we get

$$W_{\text{pol}} = \frac{1}{2}\epsilon_0 \chi^{(1)} E^2. \tag{A.8}$$

Equation (A.8) leads to an important conclusion that dielectric materials store electrostatic energy when they are polarized under external electric field E.

A.3.   Maxwell Stresses [4, 5]

A.3.1.   *Electric forces perpendicular to surface of electrodes*

The fact that dielectric polarizations generate electrostatic energy leads to another important conclusion; *dielectric polarizations are sources of mechanical stresses*. To show this, let us treat a capacitor that has electric charges $+Q$ and $-Q$ on electrodes, meanwhile the external circuit is open. We tentatively displace the electrode that has electric charges $+Q$ by a small distance δd in the direction perpendicular to the electrode surface (see Fig. A.2(a)). Experimentally, the dielectric material between the two electrodes must be mechanically strained or peel electrodes from the dielectric material to change the distance between the two electrodes, but we here think of a *virtual* displacement; we here treat dielectric materials as a space medium of dielectric constant ϵ and "imagine" a displacement without mechanical strains in these materials [4,5]. Note that more precise treatments that consider mechanical strains in dielectric materials are shown in

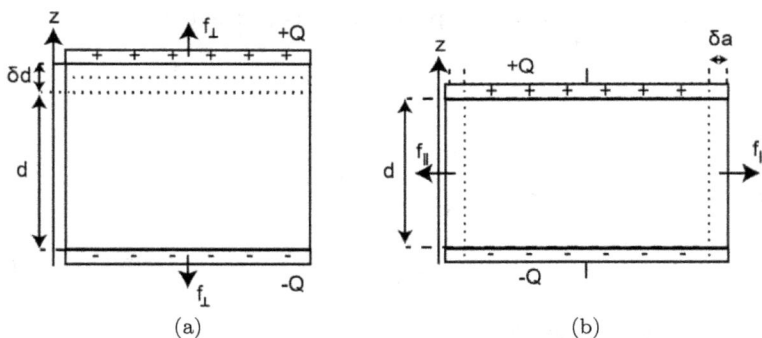

(a)   (b)

Fig. A.2:   The two electrodes of a capacitor are connected with a battery that supplies voltage V until these two electrodes store electric charges $+Q$ and $-Q$. Then, the external circuit is open, (a) and (b). For the cases that the electrode (that stores electric charges $+Q$) is displaced by δd to the z-direction, electrostatic energy stored in this capacitor changes by $\delta W_\perp$; to displace the electrode, mechanical work $f_\perp \delta d$ is necessary (a), $f_\perp < 0$. For the cases where the area of the two electrodes is increased by $\delta A = 2\pi a \delta a$, electrostatic energy stored in this capacitor changes by $\delta W_\parallel$; to change the area of the two electrodes, mechanical work $f_\parallel \delta a$ is necessary (b), $f_\parallel > 0$.

the book by Landau and Lifshitz [6], but we do not follow the precise treatments here. With the virtual displacement, the electrostatic energy of the capacitor increases by

$$\delta W_\perp = \frac{1}{2}Q^2\frac{1}{\epsilon A}\delta d. \tag{A.9}$$

In other words, the volume of the medium virtually increases from Ad to $Ad + A\delta d$, and the electrostatic energy $\delta W_\perp$ is additionally stored in the increased volume $A\delta d$. Noteworthy that the electric charges stored in the two electrodes are fixed during this *virtual* displacement because the external circuit is electrically open (see Fig. A.2(a)).

For the cases where the displacement δd is by external forces $f_\perp$ applied to the electrode, the work done by the external forces, $-f_\perp \delta d$, are equal to the changes $\delta W_\perp$ of the electrostatic energy of the system. This argument implies that electric forces $f_\perp$ that have the form

$$f_- = -\frac{1}{2}\frac{Q^2}{\epsilon A}. \tag{A.10}$$

These are indeed forces applied to the electric charges $+Q$ on the electrode due to the electric fields $Q/2\epsilon$ generated by the electric charges $-Q$ induced on the other electrode. Forces per unit area that are applied to the electrode have the form

$$\frac{f_\perp}{A} = -\frac{1}{2}\epsilon E^2, \tag{A.11}$$

and are functions of electric fields E at the space between the two electrodes. The latter fact leads to the conclusion that these electric forces result from mechanical stresses that are generated in the space between the two electrodes, where these mechanical stresses have the form

$$T_- = \frac{1}{2}\epsilon E^2. \tag{A.12}$$

We here tentatively treat mechanical stresses as forces per unit area that are applied to the surfaces of materials (in our case, the dielectric material between the two electrodes, in which electric fields are generated), whereas the mechanisms that generate these forces are operated in the bulk of these materials (mechanical stresses are thus generated in the bulk of the dielectric material). Indeed, this treatment is only effective for the cases that stresses are uniform in materials.

A.3.2.   *Electric forces parallel to surface of electrodes*

Similarly, the electric forces that are perpendicular to the direction of applied electric fields are derived: We treat a capacitor that has circular electrodes of radius a and consider a *virtual* displacement with respect to the area of these electrodes (see Fig. A.2(b)). This displacement is, again, a *virtual* displacement and thus we do not treat mechanical strains that are generated in the two electrodes and the dielectric material (between the two electrodes) by this displacement. With a virtual displacement with respect to the radii of the electrodes by δa (and thus the areas of these electrodes increase by $2\pi a \delta a$), see Fig. A.2(b), electrostatic energy that is generated in this capacitor increases by

$$\delta W_\parallel = -Q^2 \frac{d}{\epsilon A} \frac{\delta a}{a}. \tag{A.13}$$

In the cases that this (virtual) displacement is due to external forces $f_\parallel$ applied to the dielectric material (and the electrodes), mechanical works — $f_\parallel \delta a$ done by these forces are equal to the changes in the electrostatic energy. This leads to the fact that electrostatic forces $f_\parallel$ are applied to the dielectric material (and the electrodes), where these forces have the form

$$\frac{f_\parallel}{2\pi a d} = \frac{1}{2}\epsilon E^2. \tag{A.14}$$

Mechanical stresses that are generated in the dielectric material have the form

$$T_\parallel = -\frac{f_\parallel}{2\pi a d} = -\frac{1}{2}\epsilon E^2. \tag{A.15}$$

A.3.3.   *Maxwell stress*

Mechanical stresses, $T_\perp$ and $T_\parallel$, are generated at the positions in dielectric materials, where electric fields are generated, see Eqs. (A.12) and (A.15). These electrostatic stresses are called

Maxwell stresses. We thus conclude that generating dielectric polarizations in isotropic dielectric materials increases electrostatic energy and leads to the generation of Maxwell stresses.

A.4. Microscopic Mechanisms of Dielectric Polarizations [7, 8]

A.4.1. *Dielectric polarizations and electric dipoles*

The source of dielectric polarizations is electric dipoles. Electric dipoles are modeled using a pair of positive and negative charges, $+q_m$ and $-q_m$, that are separated by a small distance l (see Fig. A.3(a)). The electric dipole μ is a vector quantity that is expressed by the magnitude $q_m l$ (called electric dipole moment) and the direction defined to be parallel to a unit vector pointing from the negative charge to the positive charge. Organic molecules with (centro-) symmetric atomic arrangements do not have electric dipoles, but can show electric dipoles when electric fields are applied. These electric dipoles are called induced dipoles, and they are directed toward the applied electric fields. By contrast, organic

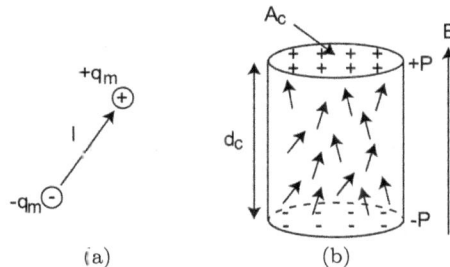

(a)   (b)

Fig. A.3:   Dielectric polarizations are due to the fact that molecules show electric dipoles that are pairs of positive and negative charges, $+q_m$ and $-q_m$, that are separated by a small distance l; the center of the gravity of the negative charges of their electrons is displaced from the center of the gravity of the positive charges of their nuclei (a). Some organic molecules that show (centro-)symmetric atomic arrangements show electric dipoles only when electric fields are applied, but many organic molecules have asymmetric atomic arrangements and thus show electric dipoles even without applied electric fields (permanent dipoles). A simple argument shows that dielectric polarizations are indeed *net* electric dipoles per unit volume (b).

molecules with asymmetric atomic arrangements have permanent electric dipoles.

To derive the relationship between macroscopic dielectric polarizations and microscopic electric dipoles, we treat a cylinder of dielectric material that has the base area Ac and height dc (see Fig. A.3). This dielectric material is composed of N_0 molecules with an electric dipole μ. Dielectric polarizations induce electric charges, $+P$ and $-P$, per unit area at the two bases; this cylinder can be viewed as an "effective" electric dipole, where electric charges, $+PA_c$ and $-PA_c$, are separated by a distance d_c. This effective electric dipole is contributed by the electric dipoles μ of N_0 constituent molecules. For the simple case that all electric dipoles are pointing toward applied electric fields, $PA_c d_c = N_0 \mu$. In general, electric dipoles are spatially distributed, and dielectric polarizations $\mathbf{P}$ are given by the vector sum of electric dipoles of N_0 constituent molecules as

$$\mathbf{P} = N_m \langle \mu \rangle, \tag{A.16}$$

where N_m $(= N_0/(Sd))$ is the volume density of electric dipoles and $\langle \mu \rangle$ is thermodynamic average with consideration of finite (orientational) distribution of electric dipoles. Dielectric polarizations are thus "net" electric dipoles per unit volume (where "net" implies that two antiparallel electric dipoles do not contribute to dielectric polarizations). Dielectric polarizations are thus vector quantities that are specified by their magnitudes and orientations.

In the case of isotropic dielectric bulk materials, $\langle \mu \rangle = 0$ in the absence of external electric fields. To generate electric dipoles, applying external electric fields is necessary. On the other hand, in the case of organic monolayers, $\langle \mu \rangle$ can be non-zero owing to the restricted orientational motion of molecules on surfaces.

A.4.2. *Microscopic polarizations of isotropic dielectric materials [7, 8]*

Four fundamental mechanisms of dielectric polarizations are summarized.

A.4.2.1. *Electronic polarizations*

Electronic polarizations are the most general mechanisms; all organic molecules and inorganic systems show these polarizations.

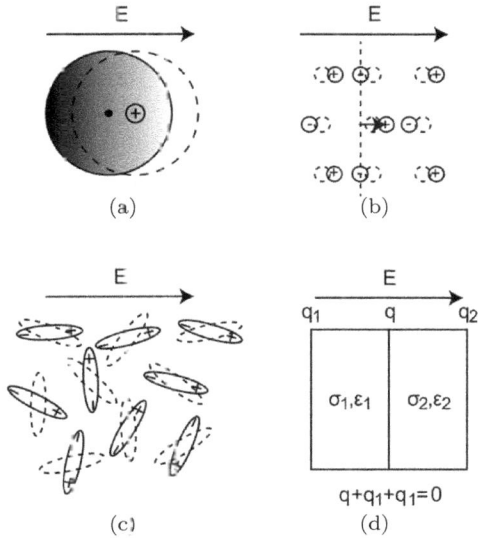

Fig. A.4: Electric dipoles are pairs of positive and negative electric charges, $+q_m$ and $-q_m$, that are separated by a small distance l. There are typical four mechanisms, where applied electric fields generate dielectric polarizations; electronic polarizations (a), ionic polarizations (b), orientational polarizations (c), and interfacial polarizations (d). The shaded region in (a) represents the distribution of electrons.

For simplicity, we here treat a symmetric molecule (the center of gravity of the negative charges of its electrons is located at the same position as the center of gravity of the positive charges of its nuclei). In applied electric fields, the central position of negative charges is displaced from that of the positive charges, see Fig. A.4(a); this displacement leads to the generation of an electric dipole, which is called an induced dipole. Electronic polarizations are indeed related to the optical properties of dielectric materials. Understanding physics of electronic polarizations is necessary for the utilizing optical instrument, e.g., Brewster–Angle microscope, instruments using optical reflection and absorption, etc.

A.4.2.2. *Ionic polarizations (atomic polarizations)*

Ionic polarizations appear in ionic crystals that are composed of positive and negative ions (see Fig. A.4(b)). In the absence of applied electric fields, constituent positive and negative ions in an ionic crystal are arranged in a periodic lattice, where the nearest neighbors of positive ions are negative ions and *vice versa*. Applied electric fields displace positive ions in the direction of applied electric fields and negative ions in the opposite direction (from their equilibrium positions). Displacing the atomic ion at each lattice point is equivalent to superposing the electric charges of electric dipoles; ionic polarizations are generated by these electric dipoles. Therefore, this polarization is also called atomic polarization.

A.4.2.3. *Orientational polarizations (dipolar polarizations)*

Organic molecules that have asymmetric atomic arrangements have permanent electric dipoles (see Fig. A.4(c)). In the absence of applied electric fields, permanent electric dipoles are oriented random and do not show dielectric polarizations; the thermodynamic average of electric dipoles is zero ($\langle \mu \rangle = 0$). In applied electric fields, permanent dipoles show orientational rotation toward the direction of applied electric fields and thus generate dielectric polarizations ($\langle \mu \rangle \neq 0$) (see Fig. A.4(c)); orientational polarizations are generated by the orientational rotation of permanent dipoles. This polarization is also called dipolar polarization.

A.4.2.4. *Interfacial polarizations (Maxwell–Wagner polarizations)*

Interfacial polarizations (or Maxwell–Wagner effects) are only generated in composite materials of lossy dielectric materials (that show dielectric properties and conduct conductive currents) or multi-layers of two or more types of lossy dielectric materials. Electrical conductions are relatively slow processes, compared with generating dielectric polarizations and thus a finite relaxation time is necessary for a dielectric material to show steady currents, where this relaxation time is determined by the ratio of conductivity and dielectric constant of this material. For the cases where two different components of a composite or a multi-layer dielectric system have different relaxation times, electric excess charges are accumulated at interfaces between these components to generate steady electric currents (see Fig. A.4(d)).

We here emphasize the fact that these accumulated charges are not bound (or polarization) charges because these charges are transported by conduction currents via lossy dielectric materials; this fact makes interfacial polarizations very different from intrinsic dielectric polarizations generated by the other mechanisms, e.g., electronic polarizations, ionic polarizations, and orientational polarizations. This polarization is one kind of space charge polarizations, where extrinsic charges contribute to the formation of interfacial space charge polarization, and it is called Maxwell–Wagner polarizations [9, 10].

In general, dielectric polarizations are generated by multiple mechanisms that are operated at the same time. A finite time is necessary to generate dielectric polarizations. Electronic polarizations are fastest processes because these are due to the displacement of electrons that have very small mass, and this polarization appears in the optical frequency region. Ionic polarizations are slower processes in comparison with electronic polarizations because these are due to the displacement of atomic ions that have thousands of times larger mass than electrons. Orientational polarizations are also slower processes in comparison with electronic polarizations because these are due to orientational rotations of permanent dipoles which are bulky in comparison with electrons, and it takes relatively long time to achieve thermodynamic equilibrium. Interfacial polarizations (Maxwell–Wagner polarizations) are even slower processes, where their characteristic time scale is given by dielectric relaxation time (conductivities divided by dielectric constants).

Dielectric processes that are slower than the frequency of (alternating) applied electric fields cannot follow the oscillation frequency of these fields and thus do not operate dielectric dispersion. The different time scale in dielectric processes allows us to separate dielectric polarization mechanisms in dielectric materials, by measuring dielectric constants as functions of the frequency of applied electric fields (*dielectric dispersion*).

References

[1]   C. Kittel, *Introduction to Solid State Physics*, 8th ed., John Wiley & Sons, Inc., New York, 2005.

[2]   M. A. Lampert and P. Mark, *Current Injection in Solids*, Academic Press, New York, 1970.

[3]   E. M. Purcell, *Electricity and Magnetism*, Berkeley Physics Course, Vol. 2, 2nd ed., McGraw-Hill Inc., New York, 1985.

[4]   R. P. Feynman, R. B. Lighton, and M. Sands, *The Feynman Lectures on Physics, Vol. II. Electromagnetism and Matter*, Basic Books, New York, 1964.

[5]   S. Sunagawa, *Theoretical Electromagnetism (Riron Denjikigaku)* (in Japanese), 3rd ed., Kinokuniya, Tokyo, 1990.

[6]   L. D. Landau, E. M. Lifshitz, and L. P. Pitaevskii, *Electrodynamics of Continuous Media*, 2nd ed., Reed Education and Professional Publishing Ltd., Oxford, 1984.

[7]   A. J. Dekker, *Solid State Physics: Electrical Engineering Materials*, Prentice-Hall, Hoboken, NJ, 1958.

[8]   T. Hino, *Electrical and Electronic Materials Property Engineering (Denkizairyou Bussei Kougaku)* (in Japanese), Asakura, Tokyo, 1985.

[9]   J. C. Maxwell, *A Treatise on Electricity and Magnetism*, Vol. 1, 3rd ed., p. 450, Dover, New York, 1954, Chap. X.

[10] S. Oka and O. Nakata, *Theory of Solid Dielectrics (Kotai Yudentai Ron)* (in Japanese), Iwanami, Tokyo, 1960.

Appendix B

Continuum Theory of Bulk Nematic Liquid Crystals

B.1. Elastic Theory of Nematic Liquid Crystals

Liquid crystal is a fluid state of materials. The discovery was made in 1888 by Friedrich Reinitzer [1]. The ordinary liquid phase is called *isotropic liquid phase* to distinguish it from *liquid crystal phase*. Liquid crystals that show liquid crystal phases at a range of temperatures between isotropic liquid and crystal phases are called thermotropic liquid crystals, and multi-component liquid crystals that show liquid crystal phases at a range of their compositions are called lyotropic liquid crystals. In general, thermotropic and lyotropic liquid crystals show a variety of liquid crystal phases. Among these liquid crystal phases, nematic liquid crystals have relatively simple structures and the physics of nematic liquid crystals provides the basis of other more complex liquid crystal phases.

In nematic liquid crystals that are comprised of rod-shaped molecules (where their positions are represented by their centers of mass and their orientations are represented by the orientations of their long-axes), the positions of these rod-shaped molecules are disordered and constituent molecules are directed *parallel* or *anti-parallel* to some extent in a direction that is specified by a unit vector **m** called *directors* (or orientational vectors). Directors **m** are thus the average orientations of constituent molecules. The extent that constituent molecules are aligned toward directors is represented using the orientational order parameter $S_2 = \langle P_2(\cos\theta) \rangle$, see Fig. 1.2

(Chapter 1). The orientational order parameters $S_1 = \langle P_1(\cos\theta) \rangle$ and $S_3 = \langle P_3(\cos\theta) \rangle$ of bulk nematic liquid crystals are zero because molecules that are directed parallel to the director are as many as molecules that are directed anti-parallel to the director on average. Because of this axial nature of orientational order, two anti-parallel director orientations, $\mathbf{m}$ and $-\mathbf{m}$, represent the same states. Many of the physical properties of nematic liquid crystals are characterized by using the orientational order parameter S_2. In the following, based on standard textbooks of the physics of liquid crystals written by de Gennes and Prost [2] and Chandrasekhar [3], the theoretical treatment of nematic liquid crystals is briefly summarized.

In bulk nematic liquid crystals, constituent molecules show long-range orientational order and thus a number of constituent molecules included in the length scale of the range of orientational ordering respond to applied fields or other perturbations *cooperatively*. Continuum theories treat the orientations of liquid crystals in this macroscopic length scale and represent their orientations by using director fields $\mathbf{m}(\mathbf{r}) = (m_x(\mathbf{r}), m_y(\mathbf{r}), m_z(\mathbf{r}))$ that are the local averages of the orientations of molecules

$$S_2(r)\left[\mathbf{m}_\alpha(\mathbf{r_0})\mathbf{m}_\beta(\mathbf{r_0}) - \frac{1}{3}\delta_{\alpha\beta}\right]$$
$$= \frac{1}{\Delta N}\int_{\Delta V} dV\left[u_\alpha(\mathbf{r}-\mathbf{r_0})u_\beta(\mathbf{r}-\mathbf{r_0}) - \frac{1}{3}|\mathbf{u}(\mathbf{r}-\mathbf{r_0})|^2\delta_{\alpha\beta}\right],$$

$$\text{(B.1)}$$

in a volume ΔV that is longer than the molecular length scale and smaller than (or as large as) the range of orientational ordering, where α and β are x, y, and z. ΔN is the number of constituent molecules in the volume ΔV and $u(\mathbf{r}) = (u_x(\mathbf{r}), u_y(\mathbf{r}), u_z(\mathbf{r}))$ are the orientations of (individual) molecules that are located at $\mathbf{r}$;

$$\Delta N = \int_{\Delta V} dV \sum_{i=1}^{N} \delta(\mathbf{r}-\mathbf{r}_i) \qquad \text{(B.2)}$$

$$\mathbf{u}(\mathbf{r}) = \sum_{i=1}^{N} \mathbf{u}_i\delta(\mathbf{r}-\mathbf{r}_i), \qquad \text{(B.3)}$$

where $\mathbf{u}_i$ is the unit vector in the orientation of the ith rod-shaped molecule, $\mathbf{r}_i$ is the positional vector of the ith rod-shaped molecule, and N is the number of molecules in the system. We cannot use a simple average of $\mathbf{u}(\mathbf{r})$ in the volume ΔV to define $\mathbf{m}(\mathbf{r})$; this average is zero for the cases of nematic liquid crystals because of the axial nature of their orientational ordering (molecules that orient towards $\mathbf{m}(\mathbf{r})$ are as many as molecules that orient $-\mathbf{m}(\mathbf{r})$ in the volume ΔV). The orientational order parameter $S_2(\mathbf{r})$ is, in general, a function of positions, but we here treat the cases that this orientational order parameter does not depend on positions because the values of the orientational order parameter are not very sensitive to applied fields or other perturbations (except at the vicinity of isotropic-nematic phase transitions).

The problems of continuum theories of nematic liquid crystals are to determine director fields $\mathbf{m}(\mathbf{r})$ for given boundary conditions and/or applied fields. The equation that determines director fields for given boundary conditions is derived by minimizing the free energy of nematic liquid crystals. In ideal cases of nematic liquid crystals, where there are no effects of boundary conditions or other perturbations (including thermal excitations of "orientational waves"), directors $\mathbf{m}$ distribute uniformly because of the fact that this is the state of minimal free energy. Boundary conditions and other small perturbations induce the gradients $\nabla \mathbf{m}(\mathbf{r})$ of directors and lead to cost-free energy. By using the fact that the volume density of free energy is invariant under the symmetry operations for the symmetry of nematic liquid crystals, the volume density of free energy arising from the gradients of directors has the form

$$f_{\mathrm{Fra}} = \frac{1}{2}k_s(\nabla \cdot \mathbf{m})^2 + \frac{1}{2}k_t(\mathbf{m} \cdot \nabla \times \mathbf{m} - t_0)^2 + \frac{1}{2}k_b(\mathbf{m} \times \nabla \times \mathbf{m})^2$$

(B.4)

with order term with respect to the gradient of directors $\mathbf{m}(\mathbf{r})$. Because Eq. (B.4) has the form of elastic energy with respect to the gradients $\mathbf{m}(\mathbf{r})$ of directors, these gradients are called orientational (elastic) deformations and Eq. (B.4) is called Frank elastic energy.

The first, second, and third terms of Eq. (B.4) are elastic energies due to splay $\nabla \cdot \mathbf{m}$, twist $\mathbf{m} \cdot \nabla \times \mathbf{m}$, and bend deformation $\mathbf{m} \times \nabla \times \mathbf{m}$, respectively (see Fig. B.1). k_s, k_t, and k_b are the Frank elastic constants for splay, twist, and bend deformations, respectively. t_0 is spontaneous twist and is zero for the cases of nematic liquid crystals that are composed of achiral molecules with no chiral dopants.

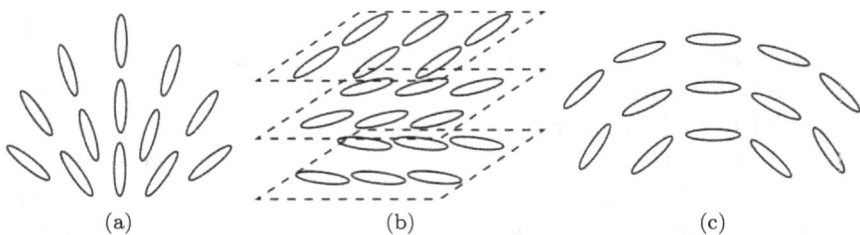

Fig. B.1:   The three modes of elastic deformations in nematic liquid crystals; splay (a), twist (b), and bend (c) deformations. The orientations $\mathbf{m}(\mathbf{r})$ of directors are parallel (or anti-parallel) to the long-axes of ellipses in the figure (this representation is because of the axial nature of nematic orientational order). Elastic energies arising from these deformations have the form of the first, second, and third terms of Eq. (B.4), respectively.

The volume integral of Eq. (B.4) all over the volume in the system is the free energy F_{Fra} of nematic liquid crystals arising from orientational deformations. Applying the variational principle to Eq. (B.4) with Lagrange multiplier $\frac{\lambda}{2}m_\alpha m_\alpha$ that ensures that the unity of the length of directors, $\mathbf{m} \cdot \mathbf{m} = m_\alpha m_\alpha = 1$, leads to Euler–Lagrange equations that have the form

$$\delta_m^{(1)} F_{\text{Fra}} = \int dS_\beta \pi_{\beta\alpha} m_\alpha(\mathbf{r}) - \int dV[h_\alpha - \lambda m_\alpha]\delta m_\alpha, \qquad (B.5)$$

with

$$h_\alpha^{\text{Fra}}(\mathbf{r}) = - \left[\frac{\partial f_{\text{Fra}}}{\partial m_\alpha} - \partial_\beta \left(\frac{\partial f_{\text{Fra}}}{\partial(\partial_\beta m_\alpha)} \right) \right], \qquad (B.6)$$

$$\pi_{\beta\alpha} = \partial_\beta \left(\frac{\partial f_{\text{Fra}}}{\partial(\partial_\beta m_\alpha)} \right), \qquad (B.7)$$

where α and β are x, y, and z, and we used the Einstein sum convention. $\partial_\beta m_\alpha$ is the partial derivative of the α-component of directors in the β-directions. The vector fields $\mathbf{h}(\mathbf{r})$ (that are conjugate to the directors $\mathbf{m}(\mathbf{r})$) are forces applied to the director at $\mathbf{r}$ due to elastic interactions with neighbor and are called "molecular fields". $\pi_{\beta\alpha}$ represent stresses arising from the elastic deformations. Because the variations $\delta m\alpha$ of directors are arbitrary small functions, the second term of Eq. (B.5) implies that the molecular fields $\mathbf{h}^{\text{Fra}}(\mathbf{r})$ are parallel to the director $\mathbf{m}$ at the position; torques arising from molecular fields are balanced at the equilibrium. The latter equation is thus called torque balance equation.

B.2. Flexoelectric Effects

Some crystals show piezoelectric effects, where dielectric polarizations are generated by applying stresses or, as converse effects, mechanical strains are generated by applying electric fields. Flexoelectric effects are analogous phenomena of liquid crystals, where dielectric polarizations are generated due to orientational deformations. This effect was first theoretically predicted by Meyer [4]. For the cases of nematic liquid crystals, dielectric polarizations arising from flexoelectric effect have the form

$$\mathbf{P}_{\text{flx}} = f_s(\nabla \cdot \mathbf{m})\mathbf{m} + f_b \mathbf{m} \times \nabla \times \mathbf{m}, \tag{B.8}$$

because of the axial symmetry of nematic liquid crystals. Dielectric polarizations are generated in nematic liquid crystals due to splay deformations (the first term of Eq. (B.8)), or due to bend deformations (the second term of Eq. (B.8)). f_s and f_b are called flexoelectric constants with respect to splay and bend deformations, and these constants are in the order of $0.01\,\text{C/m}$ for typical nematic liquid crystals, e.g., MBBA [5].

B.3. Hydrostatics

The orientational degrees of freedom of molecules in nematic liquid crystals are coupled to their translational degrees of freedom; applied shear flows drive the reorientations of directors. We here discuss the volume densities of forces and torques applied to (the centers of mass of molecules) in nematic liquid crystals due to molecular fields $h^{\text{Fra}}(\mathbf{r})$ (Eq. (B.6)). In our description of continuum theory, positional vectors $\mathbf{r}$ point to a region of volume ΔV centered at $\mathbf{r}$, where a large number of molecules are included (see Eq. (B.2)). To derive the sum of the forces $f\,\Delta V$ applied to these molecules, we use the principle of virtual displacements; we calculate the increases of the free energy due to the virtual displacements of molecules at $\mathbf{r}$ to $\mathbf{r} + \mathbf{u}(\mathbf{r})$, where their orientations $\mathbf{m}$ are fixed during these displacements; for the cases of nematic liquid crystals that show director distributions $\mathbf{m}(\mathbf{r})$, the virtual displacements change director distributions to $\mathbf{m}'(\mathbf{r}') = \mathbf{m}(\mathbf{r} - \mathbf{u}(\mathbf{r}))$. The vectors $\mathbf{u}(\mathbf{r})$ are called displacement vectors. With these virtual displacements, the variations of the gradients of directors thus have the form

$$(\partial_\beta m_\alpha) = \frac{\partial}{\partial x'_\beta} m'_\alpha(r') - \frac{\partial}{\partial x_\beta} m_\alpha(r) \simeq -\frac{\partial u_\gamma}{\partial x_\beta} \frac{\partial m_\alpha}{\partial x_\gamma}, \tag{B.9}$$

where α and β are x, y, and z. $\partial m_\alpha/\partial x_\beta$ is the partial derivative of the α-component of directors $\mathbf{m}$ in the β-direction, and $\partial m_\alpha/\partial x_\beta$ is the partial derivative of the α-component of displacement vectors $\mathbf{u}$ in the β-direction. The symbols with prime are the quantities of the distributions after the deformations.

With the Lagrange multiplier $p(r)$ that ensures the incompressibility of nematic liquid crystals, the variations of the free energy thus have the form

$$\Delta_u F_{\text{Fra}} = \int dV \pi_{\beta\alpha} \delta(\partial_\beta m_\alpha) - \int dV p(r) \nabla \cdot u(r)$$

$$= \int dV \sigma_{\beta\alpha}^{\text{Eri}} \partial_\beta u_\alpha, \tag{B.10}$$

with

$$\sigma_{\beta\alpha}^{\text{Eri}} = -p\delta_{\beta\alpha}, \tag{B.11}$$

where α and β are x, y, and z. Equation (B.10) does not include the variations of free energy due to the variations of director orientations, see Eq. (B.5). $\sigma_{\beta\alpha}^{\text{Eri}}$ is the (β, α)-component of stress tensor (called Ericksen stress tensor) applied to the molecules in the volume element at $\mathbf{r}$. It is important to notice that these stress tensors $\sigma_{\beta\alpha}^{\text{Eri}}$ are not symmetric tensors; the first subscript β indicates the direction of area elements and the second subscript is the direction of forces. The Lagrange multiplier $p(r)$ is indeed hydrostatic pressure. Net forces applied to the molecules in the volume element ΔV are $\mathbf{f}^{\text{Eri}}\Delta V = (f_x^{\text{Eri}}, f_y^{\text{Eri}}, f_z^{\text{Eri}})\Delta V$ with

$$f_\alpha^{\text{Eri}} = \partial_\beta \sigma_{\beta\alpha}^{\text{Eri}}$$

$$= -\partial_\alpha(p(r) + f_{\text{Eri}}) - h_\beta^{\text{Eri}}(r)\partial_\alpha m_\beta(r) \tag{B.12}$$

where α and β are x, y, and z. The calculation to derive the last term is slightly technical; we used Eq. (B.6) and chain rule $\partial_\gamma f_{\text{Eri}} = (\partial_\gamma n_\alpha)(\partial f_{\text{Eri}}/\partial n_\alpha) + (\partial_\gamma \partial_\beta n_\alpha)(\partial f_{\text{Eri}}/\partial(\partial_\beta n_\alpha))$, where α, β, and γ

are x, y, and z. The third term of the last equation of Eq. (B.12) is zero for the cases that molecular fields are parallel to directors at the position (because $m_\beta \partial_\alpha m_\beta = {}^1\partial_\alpha(m_\beta m_\beta) = 0$; directors are unit vectors). Equation (B.12) implies that the forces applied to nematic liquid crystals are balanced when hydrostatic pressures $p(r)$ have the form

$$p'(r) = p_0 - f_{\text{Eri}}, \qquad (B.13)$$

where p_0 is a constant.

A useful relationship is derived by using the fact that Frank elastic energy is invariant under the simultaneous rotations of positions and directors;

$$\mathbf{u(r)} = \omega \times \mathbf{r}, \qquad (B.14)$$

$$\delta \mathbf{m(r)} = \omega \times \mathbf{m(r)}. \qquad (B.15)$$

For the cases where rotational angles $|\omega|$ are small, the variations of Frank elastic energy due to the rotations, Eqs. (B.14) and (B.15), have the form

$$\delta^{(1)} F_{\text{Fra}} = \delta_m^{(1)} F_{\text{Fra}} + \delta_u^{(1)} F_{\text{Fra}}$$

$$= - \int dS_\beta \pi_{\beta\alpha} \omega_\mu \epsilon_{\mu\alpha\gamma} m_\gamma$$

$$+ \int dV [\omega_\mu \epsilon_{\mu\beta\alpha} h_\beta m_\alpha + \omega_\mu \epsilon_{\mu\beta\alpha} \sigma_{\beta\alpha}^{\text{Eri}}], \qquad (B.16)$$

where this is the sum of the contributions of the variations of director orientations, see Eq. (B.5), and displacements, see Eq. (B.10). α, β, γ, μ are x, y, and z. $\epsilon_{\mu,\alpha,\gamma}$ is so-called Levi–Civita tensor; $\epsilon_{xyz} = \epsilon_{yzx} = \epsilon_{zxy} = 1$ and $\epsilon_{yxz} = \epsilon_{xzy} = \epsilon_{zyx} = 1$ and 0 otherwise. The fact that Frank elastic energy is invariant under the operations, Eqs. (B.14) and (B.15), leads to a relationship

$$\int \epsilon_{\mu\gamma\alpha} m_\gamma (dS_\beta \pi_{\beta\alpha}) = \int dV [[\mathbf{m} \times \mathbf{h}^{\text{Fra}}]_\mu - \epsilon_{\mu\beta\alpha} \sigma_{\beta\alpha}^{\text{Eri}}]. \qquad (B.17)$$

To proceed with our calculations, we use a partial integral

$$\int dV \epsilon_{\mu\beta\alpha} \sigma_{\beta\alpha}^{\text{Eri}} = \int dV \epsilon_{\mu\beta\alpha} (\partial_\rho x_\beta) \sigma_{\rho\alpha}^{\text{Eri}}$$

$$= \int dS_\rho \epsilon_{\mu\beta\alpha} x_\beta \sigma_{\rho\alpha}^{\text{Eri}} - \int dV \epsilon_{\mu\beta\alpha} x_\beta f_\alpha^{\text{Eri}}. \qquad (B.18)$$

The conditions of force balance, Eq. (B.12), suggest that the second term of the last equation of Eq. (B.18) is zero. In this case, Eq. (B.16) is rewritten in the form

$$\delta^{(1)} F_{\text{Fra}} = \omega_\mu \int \epsilon_{\mu\gamma\alpha} m_\gamma (dS_\beta \pi_{\beta\alpha}) + \omega_\mu \int \epsilon_{\mu\gamma\alpha} x_\gamma (dS_\beta \sigma_{\beta\alpha}^{\text{Eri}})$$

$$+ \int \omega_\mu dV [h^{\text{Fra}} \times m]_\mu. \tag{B.19}$$

The volume density of torques arising from Frank elastic energy thus has the form

$$\mathbf{\Gamma}^{\text{Fra}} = \mathbf{m} \times \mathbf{h}^{\text{Fra}}. \tag{B.20}$$

The fact that molecular fields are parallel to the director at equilibrium, see the discussion in Eq. (B.5), ensures that torques are balanced in the bulk of nematic liquid crystals.

B.4.   Hydrodynamics

In applied shear flows, the directors of nematic liquid crystals show reorientations because of the coupling between the translational degrees of freedom of molecules and their orientational degrees of freedom. In the descriptions of continuum theories, we do not treat the dynamics of individual molecules, but treat collective dynamics of molecules in volume elements ΔV. In these continuum theories, the equation of translational motions of these molecules is called Navier–Stokes equation that has the form

$$\rho \frac{d}{dt} v_\alpha = \partial_\beta \sigma_{\beta\alpha}, \tag{B.21}$$

where α and β are x, y, and z; this is just the continuum description of Newton equation (the left-hand side is the time derivative of momentum per unit volume and the right-hand side is the volume density of forces). ρ is the mass density of nematic liquid crystals. $\mathbf{v}(\mathbf{r}, t) = (v_x(\mathbf{r}, t), v_y(\mathbf{r}, t), v_z(\mathbf{r}), t)$ is velocity fields in nematic liquid crystals. Navier–Stokes equations do not follow the positions of molecules as a function of time, but treat the average velocity of molecules that pass the position $\mathbf{r}$ at time t; $\mathbf{r}$ is not a dynamical variable, but just a positional vector. $\sigma_{\beta\alpha}$ is the (β, α)-component of stress tensor that has the form

$$\sigma_{\beta\alpha} = -p\delta_{\beta\alpha} + \sigma_{\beta\alpha}^{\text{Eri}} + \sigma_{\beta\alpha}, \tag{B.22}$$

where p is hydrodynamic pressures and $\delta_{\beta\alpha}$ is Kronecker's delta ($\delta_{\beta\alpha} = 1$ for $\beta = \alpha$ and 0 otherwise). In eq. (B.11), Ericksen tensor, $\sigma_{\beta\alpha}^{\text{Eri}}$ included the contributions of hydrodynamic pressures, however, we wrote these contributions separately in Eq. (B.22) to highlight that pressures are one of the unknown functions that are determined by the framework of hydrodynamics. Viscous stresses $\sigma_{\beta\alpha}^{\text{vis}}$ are functions of velocity fields $\mathbf{v}(\mathbf{r}, t)$ and director fields $\mathbf{m}(\mathbf{r}, t)$, and their functional forms depend on the properties of materials. Other stresses may add up to Eq. (B.22) for the cases of more complex materials. The equation of motion of director orientations has the form

$$I\frac{d}{dt}\left(\mathbf{m} \times \frac{d\mathbf{m}}{dt}\right) = \Gamma, \tag{B.23}$$

where I is the "inertia moment" of director and Γ is the volume density of torques that has the form

$$\Gamma = \Gamma^{\text{Fra}} + \Gamma^{\text{vis}}, \tag{B.24}$$

where Γ^{Fra} is the contribution of stresses arising from Frank elastic energy, see Eq. (B.20), and Γ^{vis} is the contribution of viscous stresses (we derive the form of Γ^{vis} in the next section). Equation (B.23) is just the continuum description of equation of rotational motions for directors (the left-hand side is the time derivative of angular momentum per unit volume and the right-hand side is the volume density of torques). The problems of the hydrodynamics of nematic liquid crystals are to derive the forms of 8 functions; $v_x(\mathbf{r}, t)$, $v_y(\mathbf{r}, t)$, $v_z(\mathbf{r}, t)$, $m_x(\mathbf{r}, t)$, $m_y(\mathbf{r}, t)$, $m_z(\mathbf{r}, t)$, $p(\mathbf{r}, t)$, and $\rho(\mathbf{r}, t)$. In this book, we only treat the cases of low Reynolds number (over damped limit), where the inertia terms (the left-hand side) of Eqs. (B.21) and (B.23) are negligible. We need two more relationships to derive the 8 unknown functions. Equations (B.21) and (B.23) are arguments by the law of the conservation of mass

$$\frac{\partial}{\partial t}\rho(\mathbf{r}, t) = -\nabla \cdot (\rho(\mathbf{r}, t)\mathbf{v}(\mathbf{r}, t)), \tag{B.25}$$

this equation ensures that the total number of molecules in the system is conserved. Many liquid materials, e.g., nematic liquid crystal, are incompressible and thus Eq. (B.25) is reduced to the form

$$\nabla \cdot \mathbf{v}(\mathbf{r}, t) = 0. \tag{B.26}$$

In these cases, mass density ρ is constant (this does not depend on both positions and time) and is not an unknown function anymore. In principle, Eqs. (B.21), (B.23), and (B.26) thus provide the forms of 7 unknown functions $\mathbf{v}(\mathbf{r}, t)$, $\mathbf{m}(\mathbf{r}, t)$, and $p(\mathbf{r}, t)$ once the functional forms of viscous stresses $\sigma_{\beta\alpha}^{\text{vis}}$ (in Eq. (B.21)) and torque density Γ (in Eq. (B.23)) are given.

B.5.   Ericksen–Leslie Viscosity and Viscous Torques

We use the phenomenological description of (linear) irreversible processes to derive the forms of viscous stresses $\sigma_{\beta\alpha}^{\text{vis}}s$ and the volume density of torques applied to directors of nematic liquid crystals. In isothermal processes (and thus the internal energy of the system does not change during the processes), the increases of entropy S_{nem} per unit time are the decrease of (Helmholtz) free energy F_{nem} per unit time;

$$T\frac{d}{dt}S_{\text{nem}} = -\frac{d}{dt}\int dV \left[\frac{1}{2}\rho v^2 + f_0 + f_{\text{Fra}}\right], \tag{B.27}$$

where $\frac{1}{2}\rho v^2$ is kinetic energy arising from flows of nematic liquid crystals, f_{Fra} is the volume density of free energy arising from orientational deformations, f_0 is the volume density of internal free energy that exists even when directors are uniform. We here use the Navier–Stokes equation, Eq. (B.22), to rewrite the time derivative of kinetic energy and calculate the time derivative of Frank elastic energy by using Eqs. (B.5) and (B.10) (both of these variations contribute to the time derivative because nematic liquid crystals show both translational and orientational motions), where $\delta m_\alpha \to \frac{d}{dt}m_\alpha$ and $\partial_\beta u_\alpha \to \partial_\beta v_\alpha$;

$$\frac{d}{dt}S_{\text{nem}} = \int dV[\sigma_{\beta\alpha}^{\text{vis}}\partial_\beta v_\alpha + h_\alpha \dot{m}_\alpha], \tag{B.28}$$

here α and β are x, y, and z. Viscous stresses σ^{vis} ($\equiv \sigma_{\beta\alpha} - \sigma^{\text{Eri}}$) are "defined" as mechanical stresses other than Ericksen stresses. $\dot{m}_\alpha$ is the time derivative of the α-component of directors that has the form

$$\dot{m}_\alpha = \frac{d}{dt}m_\alpha = \frac{\partial}{\partial t}m_\alpha(r,t) + v(r,t) \cdot \nabla m_\alpha(r,t). \qquad (B.29)$$

The first term of Eq. (B.29) is the reorientations of directors and the second term of Eq. (B.29) is because molecules that show different orientations flow into the position $\mathbf{r}$ in the next moment.

The equation of rotational motion of the system has the form

$$\left[\frac{\partial}{\partial t}\int dV \mathbf{r} \times (\rho \mathbf{v}(\mathbf{r},t))\right]_\mu$$

$$= \int \epsilon_{\mu\gamma\alpha} x_\gamma (\sigma_{\beta\alpha} dS_\beta) + \int \epsilon_{\mu\gamma\alpha} m_\gamma (\pi_{\beta\alpha} dS_\beta), \qquad (B.30)$$

where α, β, γ, μ are x, y, and x. The left-hand side of Eq. (B.30) is the time derivative of angular momentum. The first term of the right-hand side is torques with respect to translational motions and the second term of the right-hand side is torques with respect to the motions of director orientations. The left-hand side of Eq. (B.30) is rewritten by using Navier–Stokes equation, Eq. (B.21), and by integrating by parts; one of the terms is eliminated by the first term of the right-hand side of Eq. (B.30) and thus $dV\epsilon_{\mu\beta\alpha}\sigma_{\beta\alpha}$. The second term of the right-hand side of Eq. (B.30) is rewritten by using Eq. (B.17). Equation (B.30) is rewritten in the form

$$-\int dV \epsilon_{\mu\beta\alpha} \sigma_{\beta\alpha}^{\text{vis}} = \int dV [\mathbf{m} \times \mathbf{h}]_\mu, \qquad (B.31)$$

where α, β, and μ are x, y, and z. By taking into account the fact that the anti-symmetric components of stress tensors are the volume density of torques, we derive the volume of torques $\mathbf{\Gamma}^{\text{vis}} = (-\sigma_{yz}^{\text{vis}} + \sigma_{zy}^{\text{vis}}, -\sigma_{zx}^{\text{vis}} + \sigma_{xz}^{\text{vis}}, -\sigma_{xy}^{\text{vis}} + \sigma_{yx}^{\text{vis}})$ arising from viscous stresses, viscous torque, in the form

$$\mathbf{\Gamma}^{\text{vis}} = \mathbf{m} \times \mathbf{h}. \qquad (B.32)$$

By using Eq. (B.32) and the fact that any tensors are written as the sum of symmetric and anti-symmetric tensors, e.g., $\partial_\beta v_\alpha = (\partial_\beta v_\alpha + \partial_\alpha v_\beta)/2 + (\partial_\beta v_\alpha - \partial_\alpha v_\beta)/2$, Eq. (B.28) is rewritten in the form

$$T\frac{d}{dt}S_{\text{nem}} = \int dV[\sigma_{\beta\alpha}^{\text{sym}} d_{\beta\alpha} + h_\alpha M_\alpha], \qquad (B.33)$$

with

$$M = \frac{d}{dt}m - \omega \times m, \tag{B.34}$$

where α and β are x, y, and z, $\sigma_{\beta\alpha}^{\text{sym}} (\equiv (\sigma_{\beta\alpha}^{\text{vis}} + \sigma_{\alpha\beta}^{\text{vis}})/2)$ is the symmetric component of viscous stresses. $d_{\beta\alpha}$ is strain rate tensor (α and β are x, y, and z) and ω is vorticity;

$$d_{\beta\alpha} = \frac{1}{2}(\partial_\beta m_\alpha + \partial_\alpha m_\beta), \tag{B.35}$$

$$\omega = \frac{1}{2}\nabla \times v. \tag{B.36}$$

The form of Eq. (B.33) implies that the symmetric components $\sigma_{\beta\alpha}^{\text{sym}}$ of viscous stresses and molecular fields h_α are conjugate to shear rate tensor $d_{\beta\alpha}$ and M_α, respectively. The form of Eq. (B.33) implies that the symmetric components $\sigma_{\beta\alpha}^{\text{sym}}$ of viscous stresses and molecular fields h_α are conjugate to shear rate tensor $d_{\beta\alpha}$ and M_α, respectively. Onsager's formalisms of (linear) irreversible processes suggest that $\sigma_{\beta\alpha}^{\text{sym}}$ and h_α have the forms of the linear combinations of $d_{\beta\alpha}$ and M_α. Taking into account the (spatial) symmetry of the system and Onsager's reciprocal theorem, we finally derive the forms of viscous stresses $\sigma_{\beta\alpha}^{\text{sym}}$ and molecular fields h_α;

$$h_\alpha = \gamma_1 M_\alpha + \gamma_2 m_\beta d_{\beta\alpha}, \tag{B.37}$$

$$\sigma_{\beta\alpha}^{\text{vis}} = \mu_1 m_\beta m_\alpha m_\nu m_\mu d_{\nu\mu} + \mu_2 m_\beta M_\alpha + \mu_3 m_\alpha M_\beta + \mu_4 d_{\beta\alpha}$$
$$+ \mu_5 m_\beta m_\mu d_{\mu\alpha} + \mu_6 m_\alpha m_\mu d_{\mu\beta}, \tag{B.38}$$

with $\gamma_1 = \mu_3 - \mu_2$ and $\gamma_2 = \mu_2 + \mu_3 = \mu_6 - \mu_5$. In general, $\mu_2 \neq \mu_3$ and $\mu_5 \neq \mu_6$ and thus viscous stresses are not symmetric and thus generate viscous torque. $\mu_1 - \mu_6$ are so-called Ericksen–Leslie constants, see Table 4.1 (Chapter 4). Equations (B.38) and (B.38) include the coupling between the translational and orientational motions of nematic liquid crystals; translational motions of nematic liquid crystals (that are represented by strain rate tensor $d_{\beta\alpha}$) generate molecular fields **h** that drives their orientational motions, see the second term of Eq. (B.38), and, conversely, the orientational motions of nematic liquid crystals generate viscous stresses that generate their translational motions, see the second and third terms of Eq. (B.38).

References

[1] T. Geelhaar, K. Grieser, and B. Reckmann, 125 years of liquid crystals — A scientific revolution in the home, *Angew. Chem. Int. Ed.*, 52, 8798–8809 (2013).

[2] P. G. de Gennes and J. Prost, *The Physics of Liquid Crystals*, Oxford University Press, New York, 1993.

[3] S. Chandrasekhar, *Liquid Crystals*, 2nd ed., Cambridge University Press, New York, 1992.

[4] R. B. Meyer, Piezoelectric effects in liquid crystals, *Phys. Rev. Letts.*, 22, 918 (1969).

[5] Liquid crystal handbook editorial committee, *Liquid Crystal Handbook*, Maruzen, Tokyo, 2000 (in Japanese).

Appendix C

Geometric Derivation of Electric Fields Generated from Electric Dipoles

Another derivation way of the electric fields generated from electric dipoles is described, where the principle of superpositions of electrostatics is used as below. Electric fields generated from a positive electric point charge q have the form (see Fig. C.1(a)),

$$\mathbf{E}(\mathbf{r}) = \frac{1}{4\pi\epsilon_0} \frac{q}{r^2} \mathbf{e}_r, \tag{C.1}$$

where r is the distance between the position O of the positive electric point charge and the position P, at which the electric fields are generated (see Fig. C.1). $\mathbf{e}_r$ is a unit vector from O to P. Now we calculate electric fields at the position P when the positive charge is displaced by Δz, e.g., along the z-direction (see Fig. C.1(b)).

The principle of superposition suggests that displacing a positive electric charge $+q$ from O to O′ is equivalent to superposing an electric dipole that is composed of a positive electric charge at O′ and a negative electric charge at O to the positive electric charge at O (see Fig. C.1); the changes of electric fields at the position P by displacing the positive electric charge q by Δz are indeed equal to electric fields generated by an electric dipole $q\Delta z$. By the displacement, the distance between the positive charge and the point P changes to r', where $r' \simeq r - \Delta z \cos\theta'$ for $r \gg \Delta z$. This displacement thus changes

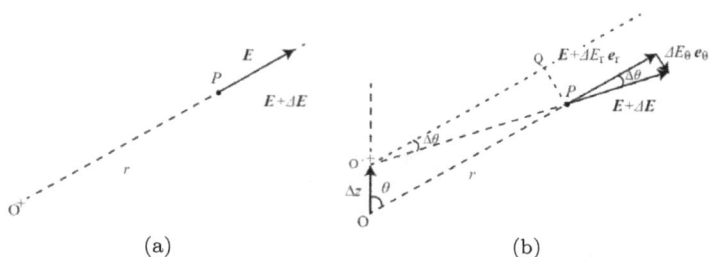

(a)   (b)

Fig. C.1:   Electric fields generated from a positive point charge q at O (a) and O′ (b). Displacing a positive charge q from O to O′ is equivalent to superposing an electric dipole (that is composed of a positive charge at O′ and a negative charge at O) to a positive charge at O. Electric fields generated from an electric dipole are thus derived as the changes of electric fields at a position P when one displaces a positive charge from O to O′.

electric fields to $\mathbf{E}'(\mathbf{r})$, where their r-components that have the form

$$E + \Delta E_r \simeq \frac{1}{4\pi\epsilon_0} \frac{q}{r'^2} = \frac{1}{4\pi\epsilon_0} \frac{q}{r^2} \left(1 + \frac{2\Delta z}{r} \cos\theta\right), \tag{C.2}$$

where E is the magnitudes of electric fields for the cases that $\Delta z = 0$ (see Eq. (C.1)). The θ-components of electric fields $E'(r)$ (that are generated by displaced electric charge) have the form

$$\Delta E_\theta \simeq E\Delta\theta, \tag{C.3}$$

where $\Delta\theta$ is the angle between these electric fields $E'(r)$ and original one $E(r)$, see Fig. C.1(b). An inspection of the geometry of the system suggests that the angle $\Delta\theta$ (between two electric fields $\mathbf{E}'(r)$ and $\mathbf{E}(r)$) is indeed equal to the angle $\angle QO'P$ in Fig. C.1(b). Because the length of the edge PQ is $\Delta z \sin\theta$ and the length of the edge O′P (that is approximately equal to the length of the edge O′Q for the cases of $r \gg \Delta z$) is r by the leading order terms with respect to $\Delta z/r$, the angle $\Delta\theta$ has the form

$$\Delta\theta \simeq \frac{\Delta z}{r} \sin\theta. \tag{C.4}$$

Substituting Eq. (C.4) into Eq. (C.3) leads to the θ-component of electric fields $\mathbf{E}'(\mathbf{r})$ in the form

$$\Delta E_\theta \simeq \frac{1}{4\pi\epsilon_0} \frac{q\Delta z}{r^3} \sin\theta. \tag{C.5}$$

Equations (C.2) and (C.5) lead to the changes of electric fields $\Delta \mathbf{E} = \Delta E_r \mathbf{e_r} + \Delta E_\theta \mathbf{e}_\theta$ by displacing the positive electric charge in the form

$$\Delta \mathbf{E} \simeq \frac{1}{4\pi\epsilon_0} \frac{2q\Delta z}{r^3} \cos\theta \mathbf{e_r} + \frac{1}{4\pi\epsilon_0} \frac{q\Delta z}{r^3} \sin\theta \mathbf{e}_\theta$$

$$\simeq \frac{1}{4\pi\epsilon_0} \left[\frac{3\mu \cdot e_r}{r^3} \mathbf{e_r} - \frac{\mu \cdot \mathbf{e}_\theta}{r^3} \mathbf{e}_\theta \right], \tag{C.6}$$

where $\mathbf{e}_\theta$ is the unit vector in the θ-direction. μ ($\equiv q\Delta z \mathbf{e}_z$) and $\mathbf{e}_z$ are the moment and direction of the electric dipole (and thus $\mu = (\mu \mathbf{e_r}) \mathbf{e_r} q \Delta z \sin\theta \mathbf{e}_\theta$, see Fig. C.1(b)). Equation (C.6) is indeed equal to Equation (2.10).

Appendix D

Taylor's Expansion for the Stability Analysis of Two-Dimensional Shape

In Taylor's expansion to the variable x, the following relations are used.

$$\mathbf{r}(s+x) = \mathbf{r}(s) + \frac{d\mathbf{r}}{ds}x + \frac{1}{2!}\frac{d^2\mathbf{r}}{ds^2}x^2 + \frac{1}{3!}\frac{d^3\mathbf{r}}{ds^3}x^3 + o(x^4)$$

$$= \mathbf{r}(s) + \mathbf{t}(s)x + \frac{1}{2}\kappa(s)\mathbf{m}(s)x^2$$

$$+ \frac{1}{6}[\kappa_s(s)\mathbf{m}(s) - \kappa(s)^2\mathbf{t}(s)]x^3 + o(x^4), \qquad (D.1)$$

and

$$\mathbf{t}(s+x) = \mathbf{t}(s) + \frac{d\mathbf{t}}{ds}x + \frac{1}{2!}\frac{d^2\mathbf{t}}{ds^2}x^2 + \frac{1}{3!}\frac{d^3\mathbf{t}}{dx^3}x^3 + o(x^4)$$

$$= \mathbf{t}(s) + \kappa(s)\mathbf{m}(s)x + \frac{1}{2}[\kappa_s(s)\mathbf{m}(s) - \kappa(s)^2\mathbf{t}(s)]x^2$$

$$+ \frac{1}{6}[(\kappa_{ss}(s) - \kappa(s)^3)\mathbf{m}(s) - 3\kappa_s(s)\kappa(s)\mathbf{t}(s)]x^3 + o(x^4),$$

$$(D.2)$$

where $\kappa_s = d\kappa(s)/ds$ and $\kappa_{ss} = d^2\kappa(s)/ds^2$. Here and in the following $o(x^n)$ refers to terms equal or higher than the nth order in x The distance and the tangential vector inner product between and at $s+x$

225

and s in the curve are calculated by

$$|\mathbf{r}(s+x) - \mathbf{r}(s)| = \left| \frac{d\mathbf{r}}{ds}x + \frac{1}{2!}\frac{d^2\mathbf{r}}{ds^2}x^2 + \frac{1}{3!}\frac{d^3\mathbf{r}}{ds^3}x^3 + o(x^4) \right|$$

$$= x \left[1 - \frac{1}{12}\kappa(s)^2 x^2 \right]^{\frac{1}{2}} + o(x^4), \qquad (D.3)$$

and

$$\mathbf{t}(s+x) \cdot \mathbf{t}(s) = 1 - \frac{1}{2}\kappa(s)^2 x^2 - \frac{3}{6}\kappa_s(s)\kappa(s)x^3 + o(x^4), \qquad (D.4)$$

Similarly, we have

$$[\mathbf{t}(s+x) \cdot \mathbf{y_0}][\mathbf{t}(s) \cdot \mathbf{y_0}]$$
$$= \sin^2 \phi(s) - \kappa(s) \cos \phi(s) \sin \phi(s)x$$
$$- \frac{1}{2}[\kappa_s(s) \cos \phi(s) \sin \phi(s) + \kappa(s)^2 \sin^2 \phi(s)]x^2$$
$$- \frac{1}{6}\{[\kappa_{ss}(s) - \kappa(s)^3] \cos \phi(s) \sin \phi(s) + 3\kappa(s)\kappa_s(s) \sin^2 \phi(s)\}x^3$$
$$+ o(x^4). \qquad (D.5)$$

We calculate $ds', \mathbf{m}', \mathbf{t}, \kappa'$ for the slightly distorted curve Eq. (D.11), e.g., the new curvature

$$\kappa' = \frac{d\mathbf{t}'}{ds'}\mathbf{m}'. \qquad (D.6)$$

Because of

$$\frac{ds'}{ds} = \left| \frac{d\mathbf{r}'}{ds} \right| = |\mathbf{t}(s) + \psi_s(s)\mathbf{m}(s) - \kappa(s)\psi(s)\mathbf{t}(s)|$$

$$= [(1 - \kappa\psi)^2 + \psi_s^2]^{\frac{1}{2}}, \qquad (D.7)$$

where $\psi_s = d\psi/ds$,

$$\mathbf{t}' = \frac{d\mathbf{r}'}{ds'} = \frac{ds}{ds'}\frac{d\mathbf{r}'}{ds} = \frac{ds}{ds'}[(1 - \kappa\psi)\mathbf{t} + \psi_s\mathbf{m}] \qquad (D.8)$$

$$\mathbf{m}' = \frac{ds}{ds'}[-\psi_s\mathbf{t} + (1 - \kappa\psi)\mathbf{m}], \qquad (D.9)$$

and $\mathbf{m}'$ is a unit vector and perpendicular to $\mathbf{t}'$, we have

$$
\kappa' = \frac{ds}{ds'} \frac{dt'}{ds} \mathbf{m}'
$$

$$
= \left(\frac{ds}{ds'}\right)^3 [(\kappa - \kappa^2 \psi + \psi_{ss})(1 - \kappa\psi) + \psi_s(\kappa_s \psi + 2\kappa\psi_s)]
$$

$$
= \kappa + \kappa^2 \psi + \psi_{ss} + o(\psi^2), \tag{D.10}
$$

where and in the following $o(\psi^n)$ refers to terms of equal or higher than the nth order in ψ. Hence the first-order variation of κ is

$$
\delta\kappa = \kappa' - \kappa = \kappa^2 \psi + \psi_{ss} + o(\psi^2). \tag{D.11}
$$

We derive other useful variations of functions with the similar processes. They are listed below:

$$
\delta(ds) = \left[-\kappa\psi + \frac{1}{2}\psi_s^2 + o(\psi^3)\right] ds.
$$

$$
\delta(\vec{r} \cdot \vec{m}ds) = [-\psi_s \vec{r} \cdot \vec{t} + (1 - \kappa \vec{r} \cdot \vec{m})\psi - \kappa\psi^2 + o(\psi^3)]ds,
$$

$$
\delta(\kappa^2 ds) = \left[\kappa^3 \psi + 2\kappa\psi_{ss} + \kappa^4 \psi^2 + \frac{3}{2}\kappa^2 \psi_s^2 + \psi_{ss}^2\right.
$$

$$
\left. + 2\kappa\kappa_s \psi_s \psi - 4\kappa^2 \psi_s s\psi + o(\psi^3)\right] ds,
$$

$$
\delta(\kappa) = \kappa^2 \psi + \psi_{ss} + o(\psi^2), \tag{D.12}
$$

$$
\delta(\kappa_s) = 3\kappa\kappa_s \psi + \kappa^2 \psi_s + \psi_{sss} + o(\psi^2),
$$

$$
\delta(\kappa_{ss}) = 3\kappa_s^2 \psi + 4\kappa\kappa_{ss} \psi + 5\kappa\kappa_s \psi_s + \kappa^2 \psi_{ss} + \psi_{ssss} + o(\psi^2).
$$

$$
\delta(\sin^2 \phi ds) = \left[-\frac{1}{2}(1 + 3\cos 2\phi)\kappa\psi + o(\psi^2)\right] ds,
$$

$$
\delta(\kappa^2 \sin^2 \phi ds) = \{[\kappa^3 (\sin^2 \phi + 2\cos 2\phi) - 4\kappa\kappa_s \sin 2\phi
$$

$$
+ 2\kappa_{ss}^2 \sin\phi]\psi + o(\psi^2)\}ds.
$$

We keep the second variations of the first three terms for the stability analysis of the domain shapes.

Appendix E

Derivation of the General Relation Concerning Shapes of Domains in Monolayers

For mathematical derivation of the **general relation** in terms of electrostatic Maxwell stress, it is necessary to use a coordinate system that represents the positions of the interior of domains that show arbitrary shapes. As a coordinate system that represents a large class of shapes of our interest, we think of a set of closed curves that are rescaled boundary curves by $t(0 < t < 1)$, see Fig. E.1, where the centers of these curves coincide with the center of the boundary curve.

Because these closed curves do not overlap each other and fill the interior of the domain, the scales t of closed curves and their contour length s uniquely determine the positions in the interior of the domain. $u(\equiv s/t)$ is the projection of a contour length s of scaled closed curves to the boundary curve and is an extension of radian to the contour length of arbitrary shapes. We use the positional vectors of the form

$$\mathbf{R}(u,t) = t\left[\mathbf{r}(u) - \frac{1}{L}\oint d\bar{s}\,\mathbf{r}(\bar{s})\right] + \mathbf{r}_0, \qquad (\text{E.1})$$

to represent the positions in the interior of the domain, where $\mathbf{r}_0(= \mathbf{R}(u,0))$ is the position of the center of the boundary curve. $\mathbf{R}(u,t)$ corresponds to the boundary curve $\mathbf{r(s)}$ and this determines $\mathbf{r_0}$ in the form

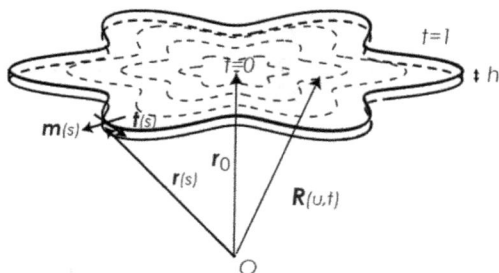

Fig. E.1: A coordinate system to represent the interior and boundary of a domain in monolayers. The interior of this domain can be filled with closed curves that are rescaled boundary curves by $t(0 < t < 1)$ without overlapping each other; t and the contour length s of rescaled curves uniquely represent the positions in the interior of domains. $u(\equiv s/t)$ is the projection of the contour length s to the boundary curve and is an extension of radian. We thus use u and t as two parameters to specify the positions in the interior of this domain.

$$\mathbf{r_0} = \frac{1}{L} \oint d\bar{s}\mathbf{r}(\bar{s}). \tag{E.2}$$

The framework of differential geometry leads to two tangent vectors that have the forms

$$\mathbf{R}_u \equiv \frac{\partial}{\partial u}\mathbf{R}(u, t) = t\mathbf{t}(u), \tag{E.3}$$

$$\mathbf{R}_t \equiv \frac{\partial}{\partial t}\mathbf{R}(u, t) = \mathbf{r}(u) - \frac{1}{L} \oint d\bar{s}\mathbf{r}(\bar{s}). \tag{E.4}$$

The coefficients of the first fundamental form thus have the form

$$g_{uu} \equiv \mathbf{R}_u \cdot \mathbf{R}_u = t^2, \tag{E.5}$$

$$g_{ut} \equiv \mathbf{R}_u \cdot \mathbf{R}_t = t\mathbf{R}_t \cdot \mathbf{t}(u), \tag{E.6}$$

$$g_{tt} \equiv \mathbf{R}_t \cdot \mathbf{R}_t. \tag{E.7}$$

These coefficients lead to the form of area elements

$$dS \equiv \sqrt{g}dudt = t\mathbf{R}_t \cdot \mathbf{m}(u), \tag{E.8}$$

where $g(\equiv g_{uu}g_{tt} - g_{ut}^2)$ is the determinant of the coefficients of the first fundamental form. This coordinate system is a natural extension of $\mathbf{r}(s)$ and thus can be used for many homotopic closed curves.

Moreover, an expansion with respect to the curvature of the boundary curve by using Frenet–Serret theorem is naturally extended to 2D space (we use this to expand the Frank elastic coefficient and anisotropic line tensions, see Eqs. (6.25), (6.26), and (6.28). With a small variation $\Gamma(s)$, the boundary curve $\mathbf{r}(s)$ of this domain changes to the form

$$\mathbf{r}'(s') = \mathbf{r}(s) + \Gamma(s)\mathbf{m}(s). \tag{E.9}$$

We use Eq. (E.1) (the parameter u in Eq. (E.1) is just an arbitrary parameter of curves) to derive the contour length of the new shape

$$ds' = (1 - \kappa(s)\Gamma(s) + O(\Gamma^2))ds. \tag{E.10}$$

Substituting Eq. (E.9) into Eqs. (E.2) and (E.4) lead to the tangent and normal vectors of the boundary curve in the forms

$$\mathbf{t}'(s') = \mathbf{t}(s) + \Gamma_s\mathbf{m}(s) + \mathbf{O}(\Gamma^2), \tag{E.11}$$

$$\mathbf{m}'(s') = \mathbf{m}(s) - \Gamma_s\mathbf{t}(s) + \mathbf{O}(\Gamma^2). \tag{E.12}$$

Equation (E.3) leads to the curvatures of the boundary curve in the form

$$\kappa'(s') = \kappa(s) + \kappa^2(s)\Gamma(s) + \Gamma_{ss}(s) + O(\Gamma^2). \tag{E.13}$$

By using Eq. (E.10), the first variation of the second term of Eq. (6.36) has the form

$$\delta^{(1)}[\lambda_0 \oint ds] \equiv \lambda_0[\oint ds' - \oint ds] = -\lambda_0 \oint ds\kappa(s)\Gamma(s). \tag{E.14}$$

With the variation $\Gamma(s)$ of the boundary curve, Eq. (E.9), the positional vectors in the interior of this domain, Eq. (E.1), change to the form

$$\mathbf{R}'(u',t') = t\mathbf{r}(u) + t\Gamma(u)\mathbf{m}(u) + \frac{1-t}{L} \oint d\bar{s}\,\mathbf{r}(\bar{s}) - (1-t)\Delta\mathbf{r}, \tag{E.15}$$

with

$$\Delta\mathbf{r} = \frac{1}{L} \oint d\bar{s}\kappa(\bar{s})\Gamma(\bar{s}) \left(\mathbf{r}(\bar{s}) - \frac{1}{L} \oint dv\mathbf{r}(v) - \frac{\mathbf{m}(\bar{s})}{\kappa(\bar{s})} \right). \tag{E.16}$$

$\Delta\mathbf{r}$ is the translation of the center of the boundary curve and we choose $\Gamma(s)$ that does not change the center of the boundary curve

$\Delta \mathbf{r} = 0$. We use new parameters u' and t' to represent the positions in the interior of this domain and the elements of these parameters have the forms

$$t' du' = \left| \frac{\partial}{\partial u} \mathbf{R}'(u', t') \right| du = t(1 - \kappa \Gamma + \mathbf{O}(\Gamma^2)) du, \qquad \text{(E.17)}$$

$$dt' = dt. \qquad \text{(E.18)}$$

The two tangent vectors of the new coordinate system read

$$\mathbf{R}_{u'}(u', t') \equiv \frac{\partial}{\partial u'} \mathbf{R}'(u', t') = t(\mathbf{t}(u) + \Gamma_u(u) \mathbf{m}(u) + \mathbf{O}(\Gamma^2)), \qquad \text{(E.19)}$$

$$\mathbf{R}_{t'}(u', t') \equiv \frac{\partial}{\partial t'} \mathbf{R}'(u', t') = \mathbf{R}_t + \Gamma(u) \mathbf{m}(u) + \mathbf{O}(\Gamma^2). \qquad \text{(E.20)}$$

The coefficients of the first fundamental form thus are calculated in the forms

$$g'_{u'u'} = t^2 + O(\Gamma^2), \qquad \text{(E.21)}$$

$$g'_{u't'} = t\mathbf{R}_t \cdot \mathbf{t}(u) + t\Gamma_u \mathbf{R}_t \cdot \mathbf{m}(u) + O(\Gamma^2), \qquad \text{(E.22)}$$

$$g'_{t't'} = \mathbf{R}_t \cdot \mathbf{R}_t + 2\Gamma \mathbf{R}_t \cdot \mathbf{m}(u) + \mathbf{O}(\Gamma^2). \qquad \text{(E.23)}$$

These coefficients of the first fundamental form lead to area elements in the forms

$$dS' \equiv \sqrt{g'} \, du' dt'$$

$$= t \left[\mathbf{R}_t \cdot m(u) + 2\Gamma(u) - \frac{d}{du}(\Gamma(u) \mathbf{R}_t \cdot t(u)) + O(\Gamma^2) \right] du dt, \qquad \text{(E.24)}$$

where $g' (\equiv g'_{u'u'} g'_{t't'} - g'^2_{u't'})$ is the determinant of the coefficients of the first fundamental form. The first order variation of the first term of Eq. (6.36) thus is derived in the form

$$\delta^{(1)} \left[\Delta P \int dS \right] \equiv \Delta P \left[\int dS' - \int dS \right]$$

$$= \Delta P \oint du \int dt t \left[2\Gamma(u) - \frac{d}{du}(\Gamma(u) \mathbf{R}_t \cdot \mathbf{t}(u)) \right]$$

$$= \Delta P \oint ds \Gamma(s), \qquad \text{(E.25)}$$

where integrating the second term of the second equation with respect to u for along a closed curve leads to zero and we integrated the first term of this equation with respect to t explicitly to derive the last equation. We also used the fact that u is equal to s for $t = 1$ to derive the last equation cf Eq. (E.25). The first-order variation of the third term of Eq. (6.36) reads

$$\delta^{(1)} F_\perp$$

$$= -P_{0\perp}^2 \int ds_i \int ds_j \left(-\Gamma(s_i) \frac{\mathbf{t}(s_i) \cdot \mathbf{t}(s_j) \mathbf{m}(s_i) \cdot (\mathbf{r}(s_i) - \mathbf{r}(s_j))}{|\mathbf{r}(s_i) - \mathbf{r}(s_j)|^3} \right)$$

$$- P_{0\perp}^2 \int ds_i \int ds_j (-\kappa(s_i)\Gamma(s_i)) \frac{\mathbf{t}(s_i) \cdot \mathbf{t}(s_j)}{|\mathbf{r}(s_i) - \mathbf{r}(s_j)|}$$

$$- P_{0\perp}^2 \int ds_i \int ds_j \Gamma_s(s_i) \frac{\mathbf{m}(s_i) \cdot \mathbf{t}(s_j)}{|\mathbf{r}(s_i) - \mathbf{r}(s_j)|}. \tag{E.26}$$

We here used the fact that $F_\perp$, the second term of Eq. (5.13), is symmetric with respect to s_i and s_j. The first, second, and third terms are the first-order variation with respect to $|\mathbf{r}(s_i) - \mathbf{r}(s_j)|$ (in the denominator of the integrand of Eq. (5.13), the second term), ds_i, and $\mathbf{t}(s_i)$ and thus used Eqs. (E.9)–(E.11), respectively. The third term of Eq. (E.26) is integrated by parts;

$$P_{0\perp}^2 \int ds_i \int ds_j \Gamma(s_i) \frac{d}{ds_i} \left(\frac{\mathbf{m}(s_i) \cdot \mathbf{t}(s_j)}{|\mathbf{r}(s_i) - \mathbf{r}(s_j)|} \right)$$

$$= -P_{0\perp}^2 \int ds_i \int ds_j \Gamma(s_i)\kappa(s_i) \frac{\mathbf{t}(s_i) \cdot \mathbf{t}(s_j)}{|\mathbf{r}(s_i) - \mathbf{r}(s_j)|}$$

$$- P_{0\perp}^2 \int ds_i \int ds_j \Gamma(s_i) \frac{\mathbf{m}(s_i) \cdot \mathbf{t}(s_j) \mathbf{t}(s_i) \cdot (\mathbf{r}(s_i) - \mathbf{r}(s_j))}{|\mathbf{r}(s_i) - \mathbf{r}(s_j)|^3}. \tag{E.27}$$

By using Eq. (E.27) for the third term of Eq. (E.26), we arrive at the form

$$\delta^{(1)} F_\perp = P_{0\perp}^2 \int ds_i \int ds_j \Gamma(s_i) \frac{\mathbf{t}(s_i) \cdot \mathbf{t}(s_j) \mathbf{m}(s_i) \cdot (\mathbf{r}(s_i) - \mathbf{r}(s_j))}{|\mathbf{r}(s_i) - \mathbf{r}(s_j)|^3}$$

$$- P_{0\perp}^2 \int ds_i \int ds_j \Gamma(s_i) \frac{\mathbf{m}(s_i) \cdot \mathbf{t}(s_j) \mathbf{t}(s_i) \cdot (\mathbf{r}(s_i) - \mathbf{r}(s_j))}{|\mathbf{r}(s_i) - \mathbf{r}(s_j)|^3}$$

$$= -P_{0\perp}^2 \int ds_i \int ds_j \Gamma(s_i)$$

$$\times \frac{(\mathbf{m}(s_i) \times \mathbf{t}(s_i)) \cdot (\mathbf{t}(s_j) \times (\mathbf{r}(s_i) - \mathbf{r}(s_j)))}{|\mathbf{r}(s_i) - \mathbf{r}(s_j)|^3}$$

$$= -P_{0\perp}^2 \int ds_i \int ds_j \Gamma(s_i) \frac{\mathbf{n} \cdot (\mathbf{t}(s_j) \times (\mathbf{r}(s_i) - \mathbf{r}(s_j)))}{|\mathbf{r}(s_i) - \mathbf{r}(s_j)|^3}.$$

$$(E.28)$$

We used a vector formula $(\mathbf{A} \times \mathbf{B}) \cdot (\mathbf{C} \times \mathbf{D}) = (\mathbf{A} \cdot \mathbf{C})(\mathbf{B} \cdot \mathbf{D}) - (\mathbf{A} \cdot \mathbf{D})(\mathbf{B} \cdot \mathbf{C})$ to derive the second term of Eq. (E.28) and used Eq. (4.4) to derive the last equation of Eq. (E.28). We use $\mathbf{M_0}$ and $\mathbf{B}(\mathbf{r}(s))$ that are defined in Eq. (E.27) to rewrite Eq. (E.28) in the form

$$\delta^{(1)} \mathbf{F}_\perp = - \oint ds \Gamma(s) \mathbf{M_0} \cdot \mathbf{B}(\mathbf{r}(s)). \qquad (E.29)$$

Induced electric charges at the interior and boundary of a domain change with the first-order variation of the shape of this domain and these changes have the form

$$\delta^{(1)} \rho_{\text{ind}}(\mathbf{R}(u,t)) \equiv \delta^{(1)} [-\nabla \cdot \mathbf{P}_{0\|}(\mathbf{R}(u,t))]$$

$$= -t\Gamma(u)\mathbf{m}(u) \cdot \nabla(\nabla \cdot \mathbf{P}_{0\|}(\mathbf{R})), \qquad (E.30)$$

$$\delta^{(1)} \sigma_{\text{ind}}(s)ds \equiv \delta^{(1)} [\mathbf{P}_{0\|}(\mathbf{r}(s)) \cdot \mathbf{m}(s)ds]$$

$$= \Gamma(s)\nabla \cdot \mathbf{P}_{0\|}(\mathbf{r}(s))ds - \frac{d}{ds}(\Gamma(s)\mathbf{P}_{0\|}(\mathbf{r}(s)) \cdot \mathbf{t}(s))ds,$$

$$(E.31)$$

where we used the fact that

$$\mathbf{m}(s) \cdot \nabla \mathbf{P}_{0\|}(\mathbf{r}(s)) \cdot \mathbf{m}(s) + \mathbf{t}(s) \cdot \nabla \mathbf{P}_{0\|}(\mathbf{r}(s)) \cdot \mathbf{t}(s) = \nabla \cdot \mathbf{P}_{0\|}(\mathbf{r}(s)).$$

$$(E.32)$$

We here note that the sum of the induced charges in the interior and boundary of the domain has the form

$$\delta^{(1)} \left[\int ds \sigma_{\text{ind}}(s) \right] = \int ds \Gamma(s) \nabla \cdot \mathbf{P}_{0\|}(\mathbf{r}(s)), \qquad (E.33)$$

$$\delta^{(1)} \left[\int dS \rho_{\text{ind}}(\mathbf{R}) \right]$$

$$= \oint du \int dt (-\nabla \cdot \mathbf{P}_{0\parallel}(\mathbf{R})) t \left[2\Gamma(u) - \frac{d}{du}(\Gamma(u)\mathbf{R}_t \cdot \mathbf{t}(u)) \right]$$

$$+ \oint du \int dt t^2 \Gamma(u) \mathbf{m}(u) \cdot \nabla(-\nabla \cdot \mathbf{P}_{0\parallel})\mathbf{R}_t \cdot \mathbf{m}(u)$$

$$= \oint du \int dt 2t\Gamma(u)(-\nabla \cdot \mathbf{P}_{0\parallel}(\mathbf{R}))$$

$$+ \oint du \int dt t^2 \Gamma(u) \mathbf{R}_t \cdot \nabla(-\nabla \cdot \mathbf{P}_0(\mathbf{R}))$$

$$= \int du \int dt \frac{\partial}{\partial t}(t^2 \Gamma(u)(-\nabla \cdot \mathbf{P}_{0\parallel}(\mathbf{R})))$$

$$= \oint ds \Gamma(s)(-\nabla \cdot \mathbf{P}_{0\parallel}(\mathbf{r}(s))). \tag{E.34}$$

The first term of the second equation of Eq. (E.34) was integrated into parts and we used the facts that

$$\frac{d}{du}(-\nabla \cdot \mathbf{P}_{0\parallel}(\mathbf{R})) = \mathbf{t}(u) \cdot \nabla(-\nabla \cdot \mathbf{P}_{0\parallel}(\mathbf{R}))$$

and

$$\mathbf{R}_t \cdot (\mathbf{t}(u)\mathbf{t}(u) + \mathbf{m}(u)\mathbf{m}(u)) \cdot \nabla = \mathbf{R}_t \cdot \nabla = \partial/\partial t,$$

to derive the third equation of Eq. (E.34). To derive $\mathbf{R}_t \cdot \nabla = \partial/\partial t$, we used the fact that the gradient ∇ indeed has the form

$$\nabla = \sum_{\alpha=u,t} \sum_{\beta=u,t} g^{\alpha\beta} \mathbf{R}_\alpha \partial_\beta, \tag{E.35}$$

with $\partial_u = \partial/\partial u$ and $\partial_t = \partial/\partial t$. $g^{uu}(\equiv g_{tt}/g)$, $g^{ut}(\equiv -g_{ut}/g)$, and $g^{tt} = (\equiv g_{uu}/g)$ are the inverse matrix of the coefficients of the first fundamental form. The first-order variation of the first term of Eq. (6.21) has the form

$$\delta^{(1)} F_\parallel^{1\text{st}} = \delta^{(1)} \left[\frac{1}{2} \int dS_i \int dS_j \frac{\nabla_i \cdot \mathbf{P}_{0\parallel}(\mathbf{R}_i)\nabla_j \cdot \mathbf{P}_{0\parallel}(\mathbf{R}_j)}{|\mathbf{R}_i - \mathbf{R}_j|} \right]$$

$$= -\int dS_j (\nabla_j \cdot \mathbf{P}_{0\parallel}(\mathbf{R}_j))\delta^{(1)} \phi_{\text{bulk}}(\mathbf{R}_j), \tag{E.36}$$

with

$$\delta^{(1)}\phi_{\text{bulk}}(\mathbf{R}_j) = -\int du_i \int dt_i 2t_i \Gamma(u_i)\frac{\nabla_i \cdot \mathbf{P}_{0\|}(\mathbf{R}_i)}{|\mathbf{R}_i - \mathbf{R}_j|}$$

$$+ \int du_i \int dt_i t_i \frac{d}{du_i}(\Gamma(u_i)\mathbf{R}_t \cdot \mathbf{t}(u_i))\frac{\nabla_i \cdot \mathbf{P}_{0\|}(\mathbf{R}_i)}{|\mathbf{R}_i - \mathbf{R}_j|}$$

$$- \int dS_i \frac{t_i \Gamma(u_i)\mathbf{m}(u_i) \cdot \nabla_i(\nabla_i \cdot \mathbf{P}_0)}{|\mathbf{R}_i - \mathbf{R}_j|}$$

$$+ \int dS_i \frac{t_i \Gamma(u_i)\mathbf{m}(u_i) \cdot (\mathbf{R}_i - \mathbf{R}_j)}{|\mathbf{R}_i - \mathbf{R}_j|^3}\nabla_i \cdot \mathbf{P}_{0\|}(\mathbf{R}_i).$$

$$(E.37)$$

We used the fact that the first term of Eq. (E.36) is symmetric with respect to the subscript i and j to derive the last equation of Eq. (E.36). The first and second terms of Eq. (E.37) are the variations arising from the area element dS_i and we used Eq. (E.24). The third term is the variation arising from the changes of induced charges at the interior of the domain, see also Eq. (E.30), and the fourth term is the variation arising from $|\mathbf{R}_i - \mathbf{R}_j|$ at the denominator of Eq. (6.21), see Eq. (E.15) The second term of Eq. (E.37) is integrated by parts;

$$-\int du_i \int dt_i t_i \Gamma(u_i)\mathbf{R}_t \cdot \mathbf{t}(u_i)\frac{\partial}{\partial u_i}\left(\frac{\nabla_i \cdot \mathbf{P}_{0\|}(\mathbf{R}_i)}{|\mathbf{R}_i - \mathbf{R}_j|}\right)$$

$$= -\int du_i \int dt_i t_i^2 \Gamma(u_i)\mathbf{R}_t \cdot \mathbf{t}(u_i)\frac{(\mathbf{t}(u_i) \cdot \nabla_i)\nabla_i \cdot \mathbf{P}_{0\|}(\mathbf{R}_i)}{|\mathbf{R}_i - \mathbf{R}_j|}$$

$$+ \int du_i \int dt_i t_i^2 \Gamma(u_i)\mathbf{R}_t \cdot \mathbf{t}(u_i)\frac{\mathbf{t}(u_i) \cdot (\mathbf{R}_i - \mathbf{R}_j)\nabla_i \cdot \mathbf{P}_{0\|}(\mathbf{R}_i)}{|\mathbf{R}_i - \mathbf{R}_j|^3},$$

$$(E.38)$$

where $\mathbf{R}_t$ in Eq. (E.38) is a function of u_i and t_i, but this fact is not explicitly written out (also in the following equations, Eqs. (E.40)–(E.42). We use Eq. (E.38) for the second term of Eq. (E.37). It is important to notice that the area element dS_i is $t_i \mathbf{R}_t \cdot \mathbf{m}(u_i)\,du_i\,dt_i$, see Eq. (E.23), and that $\mathbf{t}(u)\mathbf{t}(u)+\mathbf{m}(u)\mathbf{m}(u)$ is an identity operator;

$$\mathbf{A} \cdot (\mathbf{t}(u)\mathbf{t}(u) + \mathbf{m}(u)\mathbf{m}(u)) \cdot \mathbf{B} = \mathbf{A} \cdot \mathbf{B}, \qquad (E.39)$$

for arbitrary vectors $\mathbf{A}$ and $\mathbf{B}$. Applying these relationships to the third term of Eq. (E.37) and the first term of Eq. (E.38) and also to the fourth term of Eq. (E.37) and the second term of Eq. (E.38) leads to the form

$$\delta^{(1)}\phi_{\text{inter}}(\mathbf{R}_j) = -\int du_i \int dt_i 2t_i \Gamma(u_i) \frac{\nabla_i \cdot \mathbf{P}_{0\|}(\mathbf{R}_i)}{|\mathbf{R}_i - \mathbf{R}_j|}$$

$$+ \int du_i \int dt_i t_i^2 \Gamma(u_i) \frac{\mathbf{R}_t \cdot (\mathbf{R}_i - \mathbf{R}_j)\nabla_i \cdot \mathbf{P}_{0\|}(\mathbf{R}_i)}{|\mathbf{R}_i - \mathbf{R}_j|^3}$$

$$- \int du_i \int dt_i t_i^2 \Gamma(u_i) \frac{(\mathbf{R}_t \cdot \nabla_i)\nabla_i \cdot \mathbf{P}_{0\|}(\mathbf{R}_i)}{|\mathbf{R}_i - \mathbf{R}_j|}.$$

$$(\text{E.40})$$

By using the relationship that $\mathbf{R}_t \cdot \nabla i = \partial/\partial t_i$, see Eq. (E.35), and

$$\frac{\partial}{\partial t_r} \frac{1}{|\mathbf{R}_i - \mathbf{R}_j|} = -\frac{\mathbf{R}_t \cdot (\mathbf{R}_i - \mathbf{R}_j)}{|\mathbf{R}_i - \mathbf{R}_j|^3}. \qquad (\text{E.41})$$

Eq. (E.40) is rewritten in the form

$$\delta^{(1)}\phi_{\text{bulk}}(\mathbf{R}_j) = -\int du_i \int dt_i \frac{\partial}{\partial t_i}\left(t_i^2 \Gamma(u_i)\frac{\nabla_i \cdot \mathbf{P}_{0\|}(\mathbf{R}_i)}{|\mathbf{R}_i - \mathbf{R}_j|}\right)$$

$$= -\int ds_i \Gamma(s_i)\frac{\nabla_i \cdot \mathbf{P}_{0\|}(\mathbf{r}(s_i))}{|\mathbf{r}(s_i) - \mathbf{R}_j|}. \qquad (\text{E.42})$$

We used the fact that $u_i = s_i$ for $t_i = 1$ to derive the last equation of Eq. (E.42). Equation (E.36) thus has the form

$$\delta^{(1)}F_\|^{\text{1st}} = \int ds_i \int dS_j \Gamma(s_i)\frac{\nabla_i \cdot \mathbf{P}_{0\|}(r(s_i))\nabla_j \cdot \mathbf{P}_{0\|}(\mathbf{R}_j)}{|\mathbf{r}(s_i) - \mathbf{R}_j|}. \qquad (\text{E.43})$$

The first-order variation of the third term of Eq. (E.21) has the form

$$\delta^{(1)}F_\|^{\text{3rd}} = \delta^{(1)}\left[\frac{1}{2}\oint ds_i \oint ds_j \frac{\mathbf{P}_{0\|}(\mathbf{r}(s_i))\cdot \mathbf{m}(s_i)\mathbf{P}_{0\|}(\mathbf{r}(s_j))\cdot \mathbf{m}(s_j)}{|\mathbf{r}(s_i) - \mathbf{r}(s_j)|}\right]$$

$$= \oint ds_j \mathbf{P}_{0\|}(\mathbf{r}(s_j))\cdot \mathbf{m}(s_j)\delta^{(1)}\phi_{\text{bou}}(\mathbf{r}(s_j)), \qquad (\text{E.44})$$

with

$$\delta^{(1)}\phi_{\text{bou}}(\mathbf{R}_j) = - \oint ds_i \kappa(s_i)\Gamma(s_i)\frac{\mathbf{P}_{0\|}(\mathbf{r}(s_i)) \cdot \mathbf{m}(s_i)}{|\mathbf{r}(s_i) - \mathbf{R}_j|}$$

$$- \oint ds_i \Gamma_s(s_i)\frac{\mathbf{P}_{0\|}(\mathbf{r}(s_i)) \cdot \mathbf{t}(s_i)}{|\mathbf{r}(s_i) - \mathbf{R}_j|}$$

$$- \oint ds_i \Gamma(s_i)\frac{\mathbf{m}(s_i) \cdot (\mathbf{r}(s_i) - \mathbf{R}_j)}{|\mathbf{r}(s_i) - \mathbf{R}_j|^3}\mathbf{P}_{0\|}(\mathbf{r}(s_i)) \cdot \mathbf{m}(s_i)$$

$$+ \oint ds_i \Gamma(s_i)\frac{\mathbf{m}(s_i) \cdot \nabla_i \mathbf{P}_{0\|}(\mathbf{r}(s_i)) \cdot \mathbf{m}(s_i)}{|\mathbf{r}(s_i) - \mathbf{R}_j|}. \qquad \text{(E.45)}$$

The second term of Eq. (E.45) is integrated by parts;

$$\oint ds_i \Gamma(s_i)\frac{d}{ds_i}\left(\frac{\mathbf{P}_{0\|}(\mathbf{r}(s_i)) \cdot \mathbf{t}(s_i)}{|\mathbf{r}(s_i) - \mathbf{R}_j|}\right)$$

$$= \oint ds_i \kappa(s_i)\Gamma(s_i)\frac{\mathbf{P}_{0\|}(\mathbf{r}(s_i)) \cdot \mathbf{m}(s_i)}{|\mathbf{r}(s_i) - \mathbf{R}_j|}$$

$$- \oint ds_i \Gamma(s_i)\frac{\mathbf{P}_{0\|}(\mathbf{r}(s_i)) \cdot \mathbf{t}(s_i)\mathbf{t}(s_i) \cdot (\mathbf{r}(s_i) - \mathbf{R}_j)}{|\mathbf{r}(s_i) - \mathbf{R}_j|^3}$$

$$+ \oint ds_i \Gamma(s_i)\frac{\mathbf{t}(s_i) \cdot \nabla_i \mathbf{P}_{0\|}(\mathbf{r}(s_i)) \cdot \mathbf{t}(s_i)}{|\mathbf{r}(s_i) - \mathbf{R}_j|}. \qquad \text{(E.46)}$$

We use Eq. (E.46) for the second term of Eq. (E.45). By using the identity operator, Eq. (E.39), for the third term of Eq. (E.45)and the second term of Eq. (E.46), also for the fourth term of Eq. (E.45) and the third term of Eq. (E.45), Eq. (E.45) is rewritten in the form

$$\delta^{(1)}\phi_{\text{bou}}(\mathbf{R}_j) = \oint ds_i \Gamma(s_i)\frac{\nabla_i \cdot \mathbf{P}_{0\|}(\mathbf{r}(s_i))}{|\mathbf{r}(s_i) - \mathbf{R}_j|}$$

$$- \oint ds_i \Gamma(s_i)\frac{\mathbf{P}_{0\|}(\mathbf{r}(s_i)) \cdot (\mathbf{r}(s_i) - \mathbf{R}_j)}{|\mathbf{r}(s_i) - \mathbf{R}_j|^3}. \qquad \text{(E.47)}$$

Equation (E.44) thus has the form

$$\delta^{(1)} F_{\parallel}^{\text{3rd}} = \oint ds_i \oint ds_j \Gamma(s_i) \frac{\nabla_i \cdot \mathbf{P}_{0\parallel}(\mathbf{r}(s_i)) \mathbf{P}_{0\parallel}(\mathbf{r}(s_j)) \cdot \mathbf{m}(s_j)}{|\mathbf{r}(s_i) - \mathbf{r}(s_j)|}$$

$$- \oint ds_i \oint ds_j \Gamma(s_i)$$

$$\times \frac{\mathbf{P}_{0\parallel}(\mathbf{r}(s_i)) \cdot (\mathbf{r}(s_i) - \mathbf{r}(s_j)) \mathbf{P}_{0\parallel}(\mathbf{r}(s_j)) \cdot \mathbf{m}(s_j)}{|\mathbf{r}(s_i) - \mathbf{r}(s_j)|^3}, \qquad \text{(E.48)}$$

The first-order variation of the second term of Eq. (6.21) has the form

$$\delta^{(1)} F_{\parallel}^{\text{2nd}} = \oint ds_j \mathbf{P}_{0\parallel}(\mathbf{r}(s_j)) \cdot \mathbf{m}(s_j) \delta^{(1)} \phi_{\text{inter}}(\mathbf{r}(s_j))$$

$$+ \int dS_j (-\nabla_j \cdot \mathbf{P}_{0\parallel}(\mathbf{R}_j)) \delta^{(1)} \phi_{\text{bou}}(\mathbf{R}_j). \qquad \text{(E.49)}$$

where the forms of $\delta^{(1)} \phi_{\text{bulk}}(\mathbf{R}_j)$ and $\delta^{(1)} \phi_{\text{bou}}(\mathbf{R}_j)$ are shown in Eqs. (E.37) and (E.45). We have shown that $\delta^{(1)} \phi_{\text{bulk}}(\mathbf{R}_j)$ and $\delta^{(1)} \phi_{\text{bou}}(\mathbf{R}_j)$ are calculated in the forms of Eqs. (E.42) and (E.47), respectively. Substituting Eqs. (E.42) and (E.47) into Eq. (E.45) leads to the form

$$\delta^{(1)} F_{\parallel}^{2nd} = - \int ds_i \oint ds_j \Gamma(s_i) \frac{\nabla_i \cdot \mathbf{P}_{0\parallel}(\mathbf{r}(s_i)) \mathbf{P}_{0\parallel}(\mathbf{r}(s_j)) \cdot \mathbf{m}(s_j)}{|\mathbf{r}(s_i) - \mathbf{R}_j|}$$

$$- \oint ds_i \int dS_j \Gamma(s_i) \frac{\nabla_i \cdot \mathbf{P}_{0\parallel}(\mathbf{r}(s_i)) \nabla_j \cdot \mathbf{P}_{0\parallel}(\mathbf{R}_j)}{|\mathbf{r}(s_i) - \mathbf{R}_j|}$$

$$+ \oint ds_i \int dS_j \Gamma(s_i)$$

$$\times \frac{\mathbf{P}_{0\parallel}(\mathbf{r}(s_i)) \cdot (\mathbf{r}(s_i) - \mathbf{R}_j) \nabla_j \cdot \mathbf{P}_{0\parallel}(\mathbf{R}_j)}{|\mathbf{r}(s_i) - \mathbf{R}_j|^3}.$$

$$\text{(E.50)}$$

Equations (E.43), (E.48), and (E.50) lead to the first variation of electrostatic energy, F, arising from in-plane dipole–dipole interactions in the form

$$\delta^{(1)} F_{\parallel}$$

$$= - \oint ds_i \oint ds_j \Gamma(s_i) \frac{\mathbf{P}_{0\parallel}(\mathbf{r}(s_i)) \cdot (\mathbf{r}(s_i) - \mathbf{r}(s_j)) \mathbf{P}_{0\parallel}(\mathbf{r}(s_j)) \cdot \mathbf{m}(s_j)}{|\mathbf{r}(s_i) - \mathbf{r}(s_j)|^3}$$

$$+ \oint ds_i \int ds_j \Gamma(s_i) \frac{\mathbf{P}_{011}(\mathbf{r}(s_i)) \cdot (\mathbf{r}(s_i) - \mathbf{R}_j) \nabla_j \cdot \mathbf{P}_{01}(\mathbf{R}_j))}{|\mathbf{r}(s_i) - R_j|^3}$$

$$= - \oint ds_i \Gamma(s_i) \mathbf{P}_{0\parallel}(\mathbf{r}(s_i)) \cdot \mathbf{E}_{\parallel}(\mathbf{r}(s_i)), \qquad (E.51)$$

where $\mathbf{E}\,(\mathbf{r}(s_i))$ is electric fields arising from in-plane spontaneous polarizations $\parallel$ and has the form of Eq. (6.40). Equations (E.14), (E.25), (E.29), and (E.51) lead to the general shape equation of domains in monolayers in the form of Eq. (5.42), where this equation is equivalent to Eq. (6.41).

Appendix F

Essential Derivation Step to $\wp(\tilde{s})$ Given by Eq. (5.82)

According to the definition of the Weierstrass elliptic function [1] and the complex roots of $f(\eta)$, we have

$$s = \int_{\bar{p}}^{\infty} \{[(x-a)^2 + b^2](x+2a)\}^{-1/2}dx. \qquad (F.1)$$

Let $x - a = b/\xi$, we first transform the above equation into

$$s = \int_0^{\xi(\bar{p})} [\sqrt{(1+x^2)(b+3a\xi)\xi}]^{-1/2}d\xi. \qquad (F.2)$$

We replace $\$\xi$ by $\xi = (\zeta - c_3)/(c_3\zeta + 1)$ with $c_3 = (c^2 - 3a)/b$ and $c^4 = 9a^2+$ and do a transformation again, Eq. (F.2) then becomes

$$s = \sqrt{\frac{1 + c_3^2}{4c^2}} \int_{c_3}^{\zeta(\xi)} [(1+\zeta^2)(\zeta^2 - c_3^2)]^{-1/2}d\zeta. \qquad (F.3)$$

Let $\zeta^{-1} = \sqrt{1 - t^2}/c_3$. We then get a Jacobi's elliptic function of t as

$$2cs = \int_0^{t(\zeta)} [(1-t^2)(1 - k^2t)]^{-1/2}dt, \qquad (F.4)$$

241

with modulus $k^2 = 1/(1 + c_3^2)$. By substituting p back into t, we finally obtain equation $\wp(\tilde{s})$ given by Eq. (5.82).

References

[1]   H. Hancok, *Lecture on the Theory of Elliptic Functions*, Dover, New York, 1958.

Index